Improving Learning

Improving Learning centres on the findings from different areas of education-focused research that support evidence-informed teaching and contextualises these results to support decision-making in schools. It also describes the origins and principles of meta-analysis in education and how this identifies the successes in improving learning in classrooms. Moreover, it explains the thinking behind the 'Teaching and Learning Toolkit' and similar approaches, which seek a big-picture overview of research findings. The advantages and disadvantages of this approach are explored with practical examples. Additionally, it identifies the issues in using research evidence in education and the steps that can be taken to improve this.

It is not a manual on how to conduct a meta-analysis; instead the focus is on developing understanding of the approach in order to present its strengths and weaknesses. This understanding can advance critical engagement and effective use to improve educational outcomes for children and young people.

Steven Higgins is Professor of Education at Durham University. As a former primary school teacher, he has a particular interest in the interpretation and application of research in schools. He is the lead author of the Sutton Trust – Education Endowment Foundation Teaching and Learning Toolkit and led an Economic and Social Research Council researcher development initiative on meta-analysis.

Improving Learning

Meta-analysis of Intervention Research in Education

Steven Higgins

Durham University

CAMBRIDGE
UNIVERSITY PRESS

University Printing House, Cambridge CB2 8BS, United Kingdom

One Liberty Plaza, 20th Floor, New York, NY 10006, USA

477 Williamstown Road, Port Melbourne, VIC 3207, Australia

314-321, 3rd Floor, Plot 3, Splendor Forum, Jasola District Centre,
New Delhi – 110025, India

79 Anson Road, #06-04/06, Singapore 079906

Cambridge University Press is part of the University of Cambridge.

It furthers the University's mission by disseminating knowledge in the pursuit of
education, learning, and research at the highest international levels of excellence.

www.cambridge.org
Information on this title: www.cambridge.org/9781107033320
DOI: 10.1017/9781139519618

First published 2019

Printed and bound in Great Britain by Clays Ltd, Elcograf S.p.A.

A catalogue record for this publication is available from the British Library.

Library of Congress Cataloging-in-Publication Data
Names: Higgins, Steven, 1960– author.
Title: Improving learning : meta-analysis of intervention research in
education / Steven Higgins.
Description: Cambridge, United Kingdom ; New York, NY : University
Printing House, 2018. | Includes bibliographical references and index.
Identifiers: LCCN 2018015239 | ISBN 9781107033320 (alk. paper)
Subjects: LCSH: Education – Research – Evaluation. | Meta-analysis.
| School improvement programs.
Classification: LCC LB1028 .H46 2018 | DDC 370.72–dc23
LC record available at https://lccn.loc.gov/2018015239

ISBN 978-1-107-03332-0 Hardback

Contents

Figures

Tables

Acknowledgements

This book is about the application and interpretation of meta-analysis of intervention research in education to support teaching and learning in schools. It tells the story, often from my own personal understanding, of the development of meta-analysis and 'meta-synthesis' in particular, so as to provide a background to the thinking behind the Sutton Trust – Education Endowment Foundation Teaching and Learning Toolkit (Toolkit) and other similar approaches which seek to get a 'big picture' overview of findings from educational research using meta-analysis.

The book then presents some of the findings from the Toolkit from different areas of education research and sets these in context to support further application and use. It also lays out my personal thinking about the use of research evidence in education and some of the steps we might take to develop and improve this.

It is not a manual or a textbook about conducting a meta-analysis; there are already excellent resources to support those wishing to do this. My focus is on developing understanding of the approach so as to present its strengths and weaknesses. This will, I hope, make the limitations clearer in terms of the claims made but also help to know what can reasonably be inferred from research findings in terms of critical engagement and use.

There are many, many people who have helped me over the years in learning about meta-analysis, from Carol Fitz-Gibbon and David Moseley at Newcastle University, who provided much of the initial inspiration, and my co-researchers at the Centre for Teaching and Learning, Vivienne Baumfield and Elaine Hall in particular, who helped me with my first meta-analysis.

I learned the most about the technical aspects of meta-analysis by developing teaching materials with my colleagues, Rob Coe, Mark Newman, James Thomas and Carole Torgerson, on an Economic and Social Research Council (ESRC)-funded Researcher Development Initiative, which provided meta-analysis training for students and researchers across the social sciences. We were fortunate enough to persuade both Larry Hedges and Mark Lipsey to contribute to this programme. John Hattie has also been a very supportive colleague. At Newcastle University, our research centre invited him to speak as part of an ESRC seminar series, and he invited me to visit him in Auckland after I moved to Durham. I have benefitted hugely from his example and his advice.

The development of the Toolkit would not have been possible without the belief of Lee Elliot Major at the Sutton Trust, who took a leap of faith and funded the 'Pupil Premium Toolkit' in 2010. My research career would have been rather different without his trust and support. The Education Endowment Foundation (EEF) agreed to fund and develop the initial report into an online resource, which now includes versions for Scotland, Australia and Latin America. All of the staff at the EEF, and the Toolkit team in particular, Robbie Coleman, Danni Mason, Peter Henderson and Jonathan Kay, have provided invaluable constructive challenge and support as it has grown over the last few years.

At Durham University, thanks also go to Maria Katsipataki and Alaidde Berenice Villanueva Aguilera, who conduct much of the background work which maintains the Toolkit, and to Adetayo Kasim and ZhiMin Xiao, who have patiently supported my statistical education. Any errors or misconceptions remain my own.

A number of the chapters draw on previously published work, and I am grateful to the publishers for granting me permission to reuse these materials. These are acknowledged in each section and each chapter where appropriate.

Abbreviations

ANOVA	Analysis of variance
CI	Confidence intervals
CONSORT	Consolidated Standards for Reporting Trials
DfE	Department for Education
EEF	Education Endowment Foundation
ESRC	Economic and Social Research Council
Ofsted	Office for Standards in Education, Children's Services and Skills
PRISMA	Preferred Reporting Items for Systematic Reviews and Meta-analyses
RCT	Randomised controlled trial
SD	Standard Deviation
SE	Standard error
Toolkit	The Sutton Trust – Education Endowment Foundation Teaching and Learning Toolkit

Part I

Understanding Meta-analysis and Meta-synthesis

This part forms the first half of the book and looks at meta-analysis, its rationale (Chapter 1) and history (Chapter 2). It moves on to describe the origins and development of meta-meta-analysis, or 'meta-synthesis' (Chapter 3), which sets the background to an overview of Sutton Trust – Education Endowment Foundation Teaching and Learning Toolkit (Chapter 4). The Toolkit is an online summary of the intervention research in education for teaching 3- to 18-year-olds, designed initially to make the findings from intervention research accessible to schools and teachers in England, but increasingly used internationally.

Each chapter is headed by a number of key questions which I aim to answer in the course of the respective chapter. The writing style is a combination of academic (with citations and references) and a more reflective approach. My aim is to present key issues in evidence synthesis in education as I see them and to explain how I understand them from a personal perspective.

Chapter 1: Why Meta-analysis?
What is meta-analysis?
Why do we need it?
What can we learn from it?
What are its limitations?
Chapter 2: A Brief History of Meta-analysis
What are the origins of meta-analysis?
What can we learn from its evolution and development?
Chapter 3: Meta-synthesis in Education: What Can We Compare?
What is reasonable to compare?
How much difference do interventions make?
Chapter 4: The Teaching and Learning Toolkit
What patterns of effects appear when comparing meta-analyses?
Are there general implications for teaching and learning?

What can we infer from specific areas such as feedback and metacognition?
What do meta-analyses of digital technology and learning styles tell us?
What challenges arise in using evidence in this way?

Chapter 2 draws heavily on the *Review of Education* (Volume 4.1: 31–53[1]) article, 'Meta-synthesis and Comparative Meta-analysis of Education Research Findings: Some Risks and Benefits'. Chapter 4 has been developed from the 2016 article, 'Communicating Comparative Findings from Meta-analysis in Educational Research: Some Examples and Suggestions', which I wrote with Maria Katsipataki for the *International Journal of Research & Method in Education* (Volume 39.3: 237–254[2]), and I am grateful to the publishers John Wiley & Sons and Taylor & Francis (Informa UK Limited), respectively, for permission to reuse and develop this material.

---

[1] http://dx.doi.org/10.1002/rev3.3067
[2] http://dx.doi.org/10.1080/1743727X.2016.1166486

1   Why Meta-analysis?

Key questions
What is meta-analysis?
Why do we need it?
What can we learn from it?
What are its limitations?

Nothing Like Leather
A town fear'd a seige and held consultation
Which was the best method of fortification;
A grave skilful mason said, in his opinion,
That nothing but stone could secure the dominion;
A carpenter said, though that was well spoke,
It was better by far to defend it with oak;
A currier, wiser than both these together,
Said, 'try what you please – there's nothing like leather'.
> Daniel Fenning (1771). *The Improved Universal Spelling-Book*
> (16th Edition). London: Crowder, Baldwin & Collins, p. 36.

Introduction

This chapter presents an overview of the importance of meta-analysis in relation to our knowledge about effective teaching and learning in education. Meta-analyses of interventions about the use of phonics in reading, an area of enduring interest to policy and practice, are used to show where findings from meta-analysis of intervention research can provide a clearer picture of the complexity of findings in this particular field. These are presented through a selection of research findings from educational meta-analyses.

One of the problems that practitioners face is being presented with a vast array of research studies in education which all seem to claim they have 'significant' findings. Sometimes these are contradictory findings, such as about the effects of different ways of deploying teaching assistants; sometimes the findings are just about very different aspects of education, such as benefits of homework or of parental involvement

3

or the effects of peer-tutoring and the impact of breakfast clubs. How are busy teachers and head teachers to make sense of the conflicting information and advice? Which should they take notice of and which should be ignored? It sometimes feels like researchers are like the leather-maker in the fable at the beginning of the chapter, with each one advocating for their own particular area of interest as a solution to educational problems. This was what drew me to meta-analysis in the first place. I saw it as a way of getting an overview, of hovering above a research landscape and making sense of studies in a particular field to draw an overall conclusion about the average or typical benefits of a particular approach. It also offers a way to find out which features of different approaches are linked with the more and less successful research studies.

What Is Meta-analysis?

Meta-analysis is a statistical procedure used to combine the data or the findings from multiple studies. When the effect or the impact on learning outcomes from experimental research studies is consistent from one study to the next, meta-analysis can be used to identify this common effect by using appropriate statistical procedures to calculate a fair or unbiased average overall impact. When the effect or impact varies from one study to the next, meta-analysis can also be used to identify features of the various studies that explain the variation.

Suppose you have searched carefully and systematically and identified eight studies of the impact of peer feedback on the quality of students' writing (see Table 1.1: these are taken from a meta-analysis by Graham and colleagues published in 2015). The studies have used different measures of writing quality; they also vary in terms of design (quasi-experimental, or randomised experiments) and scale (sample size). How can we make sense of the findings overall? First, it is possible to estimate the impact using a common measure, called effect size (see the section later in this chapter for a further discussion of the strengths and weaknesses of this approach). Then we can weight each study so that it contributes fairly to an overall average (smaller studies will count less and larger studies more: there are different ways to do this, and more details are provided in Chapter 2, but the aim is to produce a fair average). This lets us summarise or synthesise the findings as shown in Table 1.1. Overall the impact across all the studies is an effect size of 0.58, which means that approximately 72% of the pupils in the peer feedback classes were above the mean of the comparison classes at the end of the different experiments (though this varied from study to study). This small

Table 1.1: *Graham's meta-analysis of peer feedback on writing quality*

Study	Design	Sample size	Effect size
Benson (1979)	QED	288	0.36
Boscolo and Ascorti (2004)	QED	122	0.97
Holliway (2004)	RCT	55	0.58
MacArthur et al. (1991)	QED	29	1.33
Olson (1990)	QED	42	0.71
Philippakos (2012)	RCT	97	0.31
Prater and Bermudez (1993)	RCT	46	0.15
Wise (1992)	QED	88	0.62
Overall		**767**	**0.58**

(given in Graham et al. 2015)

meta-analysis suggests that overall the impact of peer feedback approach has been beneficial in these research studies.

The important feature of meta-analysis is that it focuses on two questions at the same time. These are 'does it *work* (on average)?' and '*how well* does it work (on average)?' It converts findings from different studies to a common scale so that they can be combined or 'synthesised'. The value of meta-analysis becomes even more apparent the more studies you have to combine.

Statistical significance focuses on just one dimension of one of these questions in terms of whether the effect reaches a threshold relative to the scale of the experiment. Does it work in terms of the chance variability you might expect, given the sample size? I have always found the focus on statistical significance confusing, partly because of the logic of null hypothesis testing and partly because there are other things to consider in judging how convincing the findings from a single study are, which are often as, if not more, important. The use of effect sizes and meta-analysis offers a way through this confusion by identifying patterns of findings and effects across studies, providing a bigger picture than one offered by a single research study, however impressive the design and analysis.

A Common Scale to Compare Findings: Effect Size

Effect size is such an important metric that we need to examine the idea more closely. Effect size is a key measure in research generally and meta-analysis in particular. It is a way of measuring the *extent* of the difference between two groups. It is easy to calculate, easy to grasp and can be applied to any measured outcome for groups in education or in research more broadly. The value of using an effect size is that it quantifies the effectiveness of a particular intervention, relative to a comparison group. It allows us to move beyond the simplistic 'Did it work (or not)?' to the far more important 'How *well* did it work across a *range* of contexts?' It therefore supports a more systematic and rigorous approach to the accumulation of knowledge, by placing the emphasis on the most important aspect of the intervention – the size of the effect – rather than its statistical significance, which conflates the effect size and sample size. For these reasons, effect size is the most important tool in reporting and interpreting effectiveness, particularly when drawing comparisons about *relative* effectiveness of different approaches.

The basic idea is to compare groups, relative to the distribution of scores. This is the standardised mean difference between two groups. There has been some debate over the years about exactly how to calculate the effect size (see Chapter 2), but in practice most of the differences in approaches are small in the majority of contexts where effect sizes are calculated using data on pupils' learning (Xiao et al., 2016). It is important to remember that, as with many other statistics, the effect size is based on the *average* difference between two groups. It does not mean that all of the pupils will show the same improvement.

An Historical Aside: Statistical Significance

We have reached a crossroads in the history of the use of statistics in research studies. Significance testing has been used in experimental studies for nearly 100 years, but there is a growing consensus that its misuse is problematic. In 1925, Ronald Fisher (1890–1962) advocated for the idea of testing a hypothesis statistically, and named the technique 'tests of significance' in his monograph *Statistical Methods for Research Workers*. He suggested a probability of one in twenty (0.05) as a convenient cut-off level. In Fisher's approach, set out in more detail in 1935 in his book *The Design of Experiments*, you start by hypothesising the opposite of what you believe to be true, because of the problem of moving from deductive to inductive inference. This is called the *null hypothesis*: that there is no difference between the groups. You then conduct your experiment and consider how likely it is that the data that you have painstakingly collected could have occurred on the assumption that there would be no difference (which you probably didn't believe, or you wouldn't have bothered to run an experiment to find out; the double negatives add to the challenge of understanding this approach). This always seemed to me a slightly dishonest philosophical sleight of hand to avoid the problem of induction. You then set an arbitrary level of probability of between one in ten and one in a hundred, or even one in a thousand, but usually one in twenty (0.05 or 95%, as Fisher originally recommended), that you think is reasonable to draw conclusions. Depending on this level of probability, you conclude you have disproved the null. I was never convinced that one in twenty was that unlikely. The probability of getting two pairs in five-card poker or throwing a total of six with three dice in a single throw is about one in twenty.

You then make the abductive leap that the *opposite* of the null hypothesis must therefore be true and you accept the 'alternative hypothesis', which is that there really is a difference between the groups. However, it is clear that neither the null hypothesis nor its alternative has ever been proved. You have just decided that the data you have suggests that it is unlikely that there is no difference and then flipped this to conclude there really is a difference, on the basis that it (probably) won't work out like that more than one in twenty times. This decision is based on an arbitrary threshold about the likelihood of the reverse of what you actually believe. When this was first explained to me (well, actually not the first time, or even the second; it was perhaps the first time I understood what was being explained), I thought I was entering Terry Pratchett's Discworld and saw the null and its alternative balanced on the back of an infinite regress of hypothesis-bearing turtles. The use of tests of significance and null hypothesis testing certainly moved scientific inference forward in the last century. However, in education we need greater precision than the way the approach has evolved now allows. It now constrains more than it enables. This scepticism about formal hypothesis testing forms part of my advocacy for meta-analysis. I am more interested in the extent of the difference between the groups and the regularity of the extent of this difference across studies than I am about the particular level of probability of the opposite of what I really think.

Types of Effect Sizes

Effect sizes can be thought of in two broad categories. First are those that compare the extent of the average differences between two groups, relative to the spread of the scores. These are often referred to as standardised mean differences.[1] Second are those which report the relationship between two variables or the extent to which their overlap is explained,

---

[1] Such as Cohen's d, Glass's Δ or Hedges' g.

sometimes called variance-accounted for effect sizes.[2] Different effect sizes can be converted mathematically into others (a standardised mean difference (d) to a variance-accounted for effect size (r), for example). This can be done either directly, where they have a mathematical equivalence, or by estimating, based on assumptions about the distributions. In this book, the focus is on standardised mean differences, as these are most commonly used in intervention research in education. However, it is important to bear in mind that the research design from which the data is analysed, the sample or population studied, the measures used and the precise calculation method used all affect the comparability of particular effect size measures (see Appendix B for more information).

What Does an Effect Size Mean?

As an example, to help understand what an effect size means, suppose we have two classes of 25 pupils: one class is taught using an effective feedback intervention, and the other is taught as normal. The classes are performing at the same level and so are equivalent before the intervention. For this illustration, let's say that the intervention is effective with an effect size of 0.8 (which is about the average for feedback interventions). This is typically considered a 'large' effect size (Cohen, 1988). This means that the average pupil in the class receiving the feedback intervention (i.e., the one who would have been ranked 12th or 13th in their class) would now score about the same as the person ranked sixth in a control class which had not received the intervention. The rankings of the classes have moved relative to each other. This may result from a change in the outcomes for as few as 7 pupils in a class of 25.

You can also picture it this way. There are two classes of pupils walking across a field (see Figure 1.1). The pupils are walking as a mixed group, but each class are wearing matching hats, different from the other. One group are wearing striped hats, the other hats with spots. From above, the crowd is roughly circular in shape, and everyone is moving across the field in the same overall direction. A few are walking a bit faster than others and are at the front. A few stragglers are dawdling a bit behind, but the overall shape is roughly symmetrical. You can't tell if one group is further ahead than the other, but if you took a picture from above and measured the position of each child, the average would be a line across the middle of the crowd for both classes.

---

[2] Such as R-squared (R^2), eta-squared (η^2) or omega-squared (ω^2).

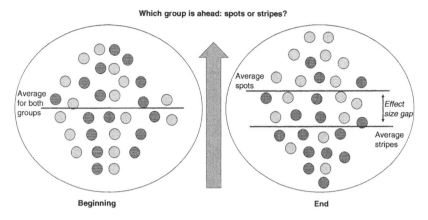

Figure 1.1: Visualising differences as effect sizes

The group with spotted hats are then given an instruction to speed up (at this point you have to imagine that the pupils in the striped-hats group don't hear this). The spotted-hats class mostly start walking a bit quicker, but this varies, as not all children react to the instruction in the same way. After ten minutes, you take another picture from above and work out the centre of each group (the average position). The spotted group have moved ahead, on average, and you can measure the distance between the two average lines (see Figure 1.1).

The effect size is a way of summarising the gap between the two classes and adjusting for how spread out each class is (a standardised effect size). Visualising the different positions of these two classes provides a helpful way of thinking about effect size. It is a measure of how far apart the two groups are as a result of the intervention (see Figure 1.2).

We can also think about what this means in a range of other ways. These can be confusing, but it is important to remember that they are all trying to express the same idea, the extent of the difference relative to the spread. Figure 1.2 displays this for an effect size, or a standardised mean difference, of 0.8.[3] For this extent of difference, 79% of the intervention group will be above the mean of the control group. This calculation is called Cohen's U3 or the percentage of pupils' scores in the lower-mean group that are exceeded by the average score in the higher-mean group. However, it is also important to note that 69% of the two groups will still overlap. There is a 71% chance that a person picked at random from the

---

[3] I am grateful to Kristoffer Magnusson for permission to use this image from his helpful website: http://rpsychologist.com/d3/cohend/.

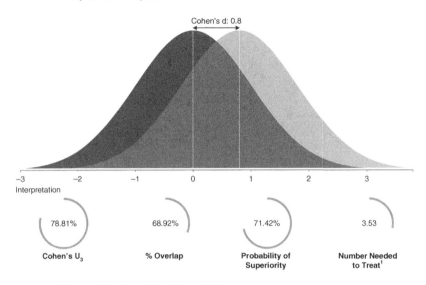

Figure 1.2: An effect size of 0.8

treatment group will have a higher score than a person picked at random from the control group (rather than 50–50 if they were the same). This is sometimes called *probability of superiority*. In terms of likely benefit, to have one more favourable outcome in the intervention group compared to the control group we would need to treat at least four pupils (technically 3.53, but I struggle with half measures here[4]). This means that if 100 pupils go through a similarly effective intervention, 28 more pupils are likely to experience a favourable outcome, compared with if they had been in the control group. This effect is typically called large, following Cohen (1988). But as Cohen himself noted, it is important to be cautious about labels like this with an arbitrary cut-off. It is worth reproducing Cohen's point in full here:

The terms 'small,' 'medium,' and 'large' are relative, not only to each other, but to the area of behavioral science or even more particularly to the specific content and research method being employed in any given investigation. In the face of this relativity, there is a certain risk inherent in offering conventional operational definitions for these terms for use in power analysis in as diverse a field of inquiry as behavioural science. This risk is nevertheless accepted in the belief that more is to be gained than lost by supplying a common conventional frame of reference

---

[4] This is the 'number needed to treat', or NNT; for those interested, see Furukawa and Leucht's (2011) paper on converting a standardised mean difference effect size (d) into the NNT.

which is recommended for use only when no better basis for estimating the ES index is available. (Cohen, 1988: p. 25)

One of the most important arguments for the approach to synthesis presented in the Toolkit (see Chapter 4) is that, whilst it is important to focus on whether a particular approach is educationally beneficial in terms of outcomes for learners, it is also important to consider others factors, such as whether it is feasible and cost-effective in relation to these effects, and to consider the quality of evidence that these conclusions are based on.

For those concerned with statistical significance, it is still readily apparent in the confidence intervals surrounding an effect size. If the confidence interval includes zero, then the effect size would be considered not to have reached conventional statistical significance (the 5% level is most commonly used in education research). The advantage of reporting effect size with a confidence interval is that it lets you judge the size of the effect first and then decide the meaning of conventional statistical significance. So, a small study with an effect size of 0.8, but with a confidence interval which includes zero, might be more interesting educationally than a much larger study with a negligible effect of 0.01, but which is statistically significant. In practice, educational interventions are rarely randomly sampled (selected at random from the population they are taken to represent), so it is not entirely clear what a significance test means in other contexts. I think of it as the *minimum* uncertainty you should allow for had those in the study been randomly sampled from a clearly defined population and sampling frame. In nearly all experimental studies in education the schools, teachers and pupils involved were selected because they were accessible or were willing participants. They are rarely a random selection from the population, and this needs to be taken into account when drawing inferences from the results.

Applying the Principles: How Important Is the Teaching of Phonics?

Let's now turn to a controversial example. How important is the teaching of phonics for beginning readers? What do we know about how effective phonics approaches are in helping young children learn to read? How secure is this evidence? What does meta-analysis tell us?

The Department of Education published a systematic review and meta-analysis by Carole Torgerson, Greg Brooks and Jill Hall in 2006. The researchers investigated how effective different phonics-based approaches were to the initial teaching of reading and spelling. The review tackled a number of specific questions:

- How effective are different approaches to phonics teaching in comparison to each other (including the specific area of analytic versus synthetic phonics[5])?
- How do different approaches impact on the application of phonics in reading and writing, including beyond the early years?
- Is there a need to differentiate by phonics for reading and phonics for spelling?
- What proportion of literacy teaching should be based on the use of phonics?

These were (and are) important questions for all those responsible for teaching reading, particularly in the first few years of formal schooling. However, phonics teaching in English-language medium schools has been a contentious issue in the teaching of literacy and the teaching of reading in particular for at least 60 years (see, for example, Rudolf Flesch's *Why Johnny Can't Read*, published in 1955, or Jeanne Chall's *Learning to Read: The Great Debate*, published in 1967). Actually, arguments about the best way to teach reading can be traced back as far as the 16th century. John Hart's (1569) *An Orthographie* and Richard Mulcaster's (1582) *Elementarie* both praised the utility of the 'alphabetic principle' through the explicit teaching of letter–sound relationships for beginning reading. In contrast, Fredrich Gedike (1754–1803) was prominent in advocating a 'whole-to-part' approach to the teaching of reading. (For specific historical details, see Davies, 1973.)

There is general agreement that *some* phonics may be needed in the teaching of early reading, but how much and of what kind has less consensus. Torgerson and colleagues' approach was to undertake a systematic search for the kinds of studies which were designed to show a clear causal link. This meant looking for controlled experiments or 'trials' where the study design included allocating the pupils to groups randomly to reduce the chance of any bias involved in choosing pupils for the intervention which might affect the analysis (such as choosing pupils that teachers think deserve the intervention is suitable for or avoiding

---

[5] In the review, 'analytic' phonics is defined as 'A form of phonics teaching in which sounding-out is not used. Instead, teachers show children how to deduce the common letter and sound in a set of words which all begin (or, later, end) with the same letter and sound, e.g. pet, park, push, pen.' 'Synthetic' phonics is defined as 'A form of phonics teaching in which sounding-out is used. For reading, this is based on the letters in printed words and is followed by blending their sounds to produce a spoken word which the learner should recognise. The classic example is "kuh" "a" "tuh" = "cat". For writing, sounding-out is based on a spoken word which the learner knows and is followed by writing the corresponding letter for each sound.'

Summary of findings, by research question, answer, quality of evidence, strength of effect, statistical significance and implications for teaching

Research question	Answer	Quality of evidence	Strength of effect	Statistical significance	Implications for teaching
Does systematic phonics instruction enable children to make better progress in reading *accuracy* than unsystematic or no phonics?	Yes*	Moderate	Small (effect size = 0.27)	Highly statistically significant (p = 0.002)	No warrant for NOT using phonics – it should be a routine part of literacy teaching
Did the evidence for the finding above differ according to whether or not researchers had used intention to teach analysis?	No	Moderate	Small (effect sizes = 0.24 and 0.37 respectively)	Not statistically significant (p>0.05) N.B. The non-significant value implies no difference between the groups.	(n/a – methodological question)
Does systematic phonics instruction enable both normally developing children and those at risk of failure to make better progress in reading *accuracy* than unsystematic or no phonics?	Yes*	Moderate	Medium and small (effect sizes = 0.45 and 0.21, respectively)	Not statistically significant (p>0.05) N.B. The non-significant value implies no difference between groups	No warrant for NOT using phonics with either group – both normally developing children and those at risk of failure can benefit
Does system phonics instruction enable children to make better progress in reading *comprehension* than unsystematic or no phonics?	Not clear	Weak	Small (effect size = 0.24)	Not statistically significant (p = 0.08)	No clear finding from research on whether or not phonics boosts progress in comprehension
Does systematic phonics instruction enable children to make better progress in spelling than unsystematic or no phonics?	Not clear	Weak	Very small (effect size = 0.09)	Not statistically significant (p = 0.56)	No warrant from research for either using or not using phonics to teach spelling
Does systematic synthetic phonics instruction enable children to make better progress in reading *accuracy* than systematic analytic phonics?	Not clear	Weak	Very small (effect size = 0.02)	Not statistically significant (p = 0.87)	No warrant from research for choosing between these varieties of systematic phonics

Figure 1.3: Summary table of findings from Torgerson et al. (2006)

those thought difficult) and to avoid a statistical problem called 'regression to the mean' where studies can appear to find a positive impact, but only because they include more very low performing pupils in the intervention group who tend to do better when tested again (they 'regress to the mean'). Random allocation typically addresses both of these issues. Their review identified a total of 20 'randomised controlled trials' (RCTs,) of which only one had been conducted in the UK. All of the studies looked at the initial teaching of reading (and, in a few cases, spelling). The children included in the studies were mostly between five and seven years of age, with four including older children up to age 11. Fourteen studies contained data which was used to draw conclusions about the effects on children's reading.

The review found that across the 14 studies that systematic phonics teaching was linked with moderately better progress in the accuracy of children's reading for the groups which received phonics teaching. This effect was seen for different levels of prior attainment, both children progressing normally and those who needed extra support. The evidence for the impact of phonics on reading comprehension was weaker. Although extent of the effect was similar (an effect size of 0.24 compared with 0.27), the finding for reading comprehension was not statistically significant, suggesting that we can't rule out the possibility it was as a result of chance statistical variation, given the available data. The review found no evidence that synthetic phonics instruction was better than analytic phonics (or vice versa) but there were only three small-scale controlled trials with randomisation from which to make this comparison (see Figure 1.3).

Random Allocation

Random allocation to groups in an experimental design is important because it allows for factors we don't or can't know in advance which might systematically affect learning in the groups being compared. We know many of the things which are likely to influence progress, such as pupils' prior attainment, their relative age, whether they have special educational needs or if they have lower attendance. However, we don't know and can't predict everything in advance.

If we randomly allocate pupils into groups (or, in a larger experiment, randomly allocate classes or even schools), then any differences we know about AND those we don't are more likely to be evenly spread between the groups. Random allocation does not guarantee this, but makes it more likely, on average. Personally, I think of randomisation as the researcher's act of humility. We accept we don't (and can't) know in advance what might systematically cause variation in effects so, where feasible and ethical, we should randomise to take account of what we don't know.

In the studies reviewed, phonics teaching did not appear to benefit children's progress in spelling, but again there were only three randomised controlled trials which looked at this. The authors concluded that this does not provide strong evidence either for or against the use of phonics in the teaching of spelling. It was not possible to analyse how different approaches affected the application of phonics in reading and writing beyond the early primary years because only three of the studies had used follow up measures. Overall the evidence summarised in this review in 2006 indicates that phonics for reading can be effective, both generally and as a catch-up for children not making effective progress and that it is a useful tool for teachers when planning the teaching of reading to primary age pupils. What is also important to note is where the evidence is lacking or inconclusive. The combination of these studies indicates that we can't yet decide whether an analytic or a synthetic approach to phonics is superior and that we can't identify if there is an optimal approach.

Summarising a Meta-analysis

One useful feature of a meta-analysis is the way that you can display your findings. My personal favourite is called a 'forest' plot, supposedly because of the 'forest' of lines[6] that result from showing each study, with its confidence interval, and the overall pooled effect, shown as a diamond, where the points indicate the overall confidence interval of this effect. It contains a large amount of information so takes some getting used to, but once you are familiar with the structure, then you can read it quickly and get a good overview of the number and size of studies and the overall spread of effects.

The structure of a forest plot follows a similar format. It contains a list of the studies, shows the impact of each study with its confidence intervals (this is sometimes repeated in numerical form) and at the bottom an overall 'pooled' average, indicating the overall findings. You can tell how much each study contributes to the overall effect by the size of the black square (this is also given as the 'weight' in the final column). (For a more detailed explanation of how to read and interpret a forest plot, using the example below, see Appendix C.) This way of displaying results

---

[6] One of the pioneers of meta-analysis in medicine, Richard Peto, is reported to have said humorously that the plot was named after a researcher called Pat Forrest, and as a result the name has sometimes been spelled 'Forrest plot'.

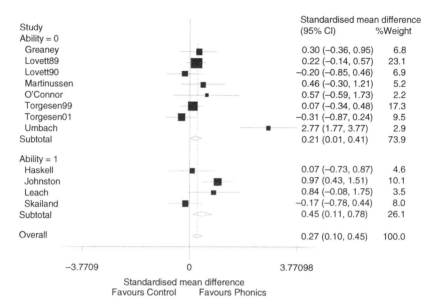

Figure 1.4: Forest plot from Torgerson et al. (2006)

is also an effective way of presenting one of the most important arguments for meta-analysis in education. If we only looked at the Lovett, Torgesen and Skailand studies we would come to the conclusion that phonics is not effective for teaching reading. Or if we only selected the research by Johnston, Umbach and Leach we might come to an overly optimistic conclusion about how helpful phonics is. It is only by looking across the available evidence and taking it all into account that we can come to a balanced conclusion.

In Torgerson and colleagues' meta-analysis, they undertook a nice exploration of phonics interventions for struggling readers (labelled as ability = 0 in Figure 1.4) and normally developing readers (labelled as ability = 1). Each of these subsets have an overall average calculated (shown as an empty diamond).

In their meta-analysis, phonics instruction tended to produce a larger effect size (0.45) for normally attaining children (the studies from Haskell downwards) than for children with reading difficulties (0.21) at the top of the list of studies. However, when they tested this interaction they could not rule out the possibility that the difference between these groups was the result of chance.

The observant may notice a study that may seem like an outlier, the Umbach study with an effect size of 2.77 (Umbach et al., 1989). This may seem incredible, but it just means that all of the children in the intervention group must have scored above the mean of the controls or about three-quarters of them (n = 15) benefited substantially from the intervention. In many meta-analyses this would be excluded as an 'outlier' or statistical anomaly. Torgerson and colleagues ran the analysis with and without this study, and it makes a difference at the level of statistical significance. The confidence interval includes zero without it, using a random effects model, but the finding would be considered conventionally statistically significant (at the 95% level) if it is retained. Some may therefore conclude the issue is not proven. However, the study by Umbach and colleagues had a particularly intensive approach to phonics where the intervention group experienced 50 minutes of the Reading Mastery approach with synthetic phonics each day for a whole school year. The comparison group had 50 minutes a day of the usual approach used in the school system. So, you could also argue the other way; that it should be included as it is an example of a highly intensive phonics programme. It is very unusual for the conclusions of a meta-analysis to hang on a single small study in this way. My conclusion would be tentatively positive, but let's get more evidence!

This is not the only meta-analysis of phonics. In the Sutton Trust – Education Endowment Foundation Teaching and Learning Toolkit's technical appendix to the Phonics entry there are eight meta-analyses (see Figure 1.5 below: the bubble plot shows recent meta-analyses of phonics with the size of the bubble proportional to the number of studies included). The pooled effect size estimates for phonics range from 0.24 to 0.47, with some of the variation explained by intensity (particularly one-to-one and small group) and outcome measures (higher effects for word level measures and lower estimates for reading comprehension). Overall the evidence is clear and consistent; in research studies phonics approaches have helped young children learn to read, but it is not a panacea (for further discussion about the evidence for teaching aspects of literacy see Chapter 6, and further details about these meta-analyses can be found in Appendix A.1.1). In terms of the interpretation of effect size presented above, with an effect size of 0.3 for a successful phonics intervention, if you gave this to 100 pupils, you would expect about 10 of them to benefit compared with the control group.[7] The evidence is substantial and robust, using a reading approach with phonics will undoubtedly help some young

---

[7] This does therefore raise a question about what the other 90 students might need to accelerate their reading. More phonics or other approaches?

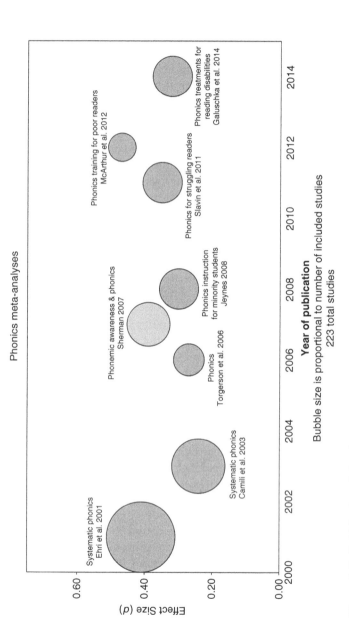

Figure 1.5: Bubble plot of phonics meta-analyses by year

readers to become more fluent and more successful as part of their reading progress.

Educational Research and Phonics from a Personal Perspective

In the UK, systematic phonics has been advocated at least since Daniels and Diack's 'Royal Road Readers'. These were first published in the 1950s, and I remember reading them myself in primary school. Twenty years later, I started as a primary school teacher at the peak of the 'real books' movement, which emphasised enjoyment and meaning in learning to read through 'authentic' texts. This approach was also based on research 'evidence'. The research base drew on the benefits of the 'Whole Language' approach, the use of natural texts and 'Big Books' and the adoption of Chomsky's 'neurological impress' method, with children listening to books on cassette tapes through headphones and following the text in a printed copy of the book (see Tunnell and Jacobs (1989) for a presentation of the research findings in literature-based reading instruction and the use of 'real books'). I lost count of the number of books I recorded as the school budget did not run to buying the commercially produced sets of books and tapes. I also recall sub-verting the 'whole language' approach with colleagues, by adding in books from rescued phonics schemes to the 'Reading Streets', creating our own not-so-royal roads which provided a structure and progression in phonic mastery. This was particularly for those children who were not making the progress we expected or hoped and who we nudged in their choices towards the somewhat stilted stories of Ben the Dog, Jip the Cat and Deb the Rat with restricted vocabulary and a clear phonics progression.[8] Teachers have always taken research evidence with a pinch of salt, but then the research community have also not system-atically addressed practitioners' concerns in terms of generating a coherent evidence base which allows comparisons to be made system-atically across programmes and approaches. When I first read Stahl and Miller's (1989) meta-analysis of whole language approaches in the late 1990s, I wished I had known about this evidence when I was a teacher of 5–6-year-olds in the East End of Newcastle upon Tyne, in the year it was published. The mean of all 117 effects is 0.09, but not reliably different from zero, indicating that whole language and

---

[8] These were characters in the 'New Way' reading series published by Nelson Thornes, which grew out of the 'Gay Way' reading books by E. R. Boyce published in 1977 by Macmillan Education.

language experience approaches may not be really different from control group approaches in their effects. This does not mean they did not 'work', just that, on average, they were no better than what teachers usually did. Children did learn to read and write, but only as well as they had before.

Chall (1967, p. 314) had concluded more than 20 years earlier:

The research in beginning reading itself needs improvement. As I have indicated throughout this book, too much of the research was undertaken to prove that one ill-defined method was better than another ill-defined method. We need, instead, series of coordinated laboratory as well as extensive longitudinal studies – studies which give us some definitive answers so we don't keep researching the same issues over and over again. I do not believe that the kind of research can be left to the schools – to teachers and administrators, who necessarily become emotionally involved with a given approach. Nor can it be left to the part-time interests of a few reading specialists. The conflict between the enthusiasm and involvement that the practitioner must have and the cool, dispassionate observation and analysis that mark the true educational experimenter will not be solved if the experimentation is left to those concerned more with helping that with knowing.

Likewise, the authors of reading programs are not the best researchers to test their own programs. While they should be involved in the early pilot studies of their programs, comparing one's own product with the products of others requires the kind of objectivity that it is difficult for any author to develop towards his [sic] own work.

My final recommendation is that experiments in beginning reading not be undertaken as if they were the first studies of their kind. Research in reading should follow the norms of science. Each researcher must try to learn from the work of those who preceded him [sic] and to add a unified body of knowledge – knowing that neither he [sic] nor anyone following him [sic] will ever have the final word.

What I found perplexing when I read the review is that the public, professional and academic debate has not moved on significantly since the 1950s. Chall's final conclusion could have been written (apart from the masculine pronouns) in 2017, 50 years later. We have not been able to accumulate and systematise the evidence from research to provide a more precise set of answers to the questions such as how much phonics and how frequently, for whom and when. For a teacher of five- to six-year-olds, how can we diagnose when to emphasise decoding and when we need to develop understanding of meaning or have a focus on language knowledge and the use of new vocabulary in learning to read? There will not be a deterministic solution for each child or class, but we should at least be able to work out the probabilities with a greater degree of certainty that we can now.

Key Principles in a Robust and Reliable Meta-analysis

One way of looking at meta-analysis is that it is like undertaking a questionnaire or a survey of the existing research. The search terms are like the sampling frame and determine what gets included. Like looking through a telescope you have to line it up effectively to be sure you are capturing what you want to look at. Once you have decided the focus the next challenge is in finding the relevant studies. There are two main approaches. You can try to be as exhaustive as you can and find everything there is. This may mean combining a range of strategies to uncover everything available. The alternative is to produce as representative a sample of studies as you can. This means thinking about what might cause bias in the field of interest and skew your findings. My worry with the exhaustive approach is that you can never know if you have found everything and that some strategies (such as citation searching) may increase the risk of some kinds of bias (with citation searching it may increase any effect of publication bias). Either way, once you have found the studies and checked they match the criteria you set for inclusion, you then have to identify the relevant information in each study and pull it out into a form you can then analyse. This feels like answering a questionnaire on behalf of each study.

The final phase of a meta-analysis as aligning the research studies so that you can look across them to see the patterns of similarity and difference. You take the information included in them that is relevant to your own inquiry and organise it to that you can compare and contrast this data effectively. The more features you want to align, the more patterns you can explore, but also the more time it will take to extract and organise the data. One typical challenge is that the underlying studies were not designed with meta-analysis in mind and you often have to make decisions about the best fit for some categories for your own research questions. Some will report the total length of time for an intervention; others will include both duration and frequency. Some are clear about what happened in the control group; others never mention this. Some will report on pupil characteristics like gender and socio-economic status; others will record proportions of special educational needs or English as an additional language. None will record or report all of the information you want, like respondents missing out questions on a survey. You can just consider the findings or the impact reported and produce an overall average to see the extent to which an approach or intervention makes a difference, but this is not very informative, particularly if there is a wide spread of effects. We want to work out whether something is effective and how great the impact is, but then it would be useful to know if longer interventions tend to work better, whether it appears more effective with older or younger learners and whether there are particular features of an approach which are linked with more successful outcomes. This is called a 'moderator analysis' as you look to see how these features correlate with or moderate the overall effect. Unless you identify or extract this information from each study, you can't address the question across studies (Chapter 7 looks at the 'moderators' across the meta-analyses of parental involvement in more detail).

What makes a good meta-analysis is hard to define precisely as some judgement is inevitably required on the part of the researchers. There are technical standards for meta-analysis set out in the PRISMA (Preferred Reporting Items for Systematic Reviews and Meta-analyses) guidelines[9] and a good meta-analysis certainly has a clear rationale and question, as well as a credible search strategy and clear inclusion criteria. The methods should be clearly described and the analysis appropriate to the question. What is harder to define is how clear the authors are in explaining what they have done and explaining the reasons why. In education, there is almost always greater variation (heterogeneity) than you would expect from the number and scale of the studies included (sampling). A good meta-analysis attempts to explain this by exploring features of the studies with a moderator analysis by looking at the associations between different features of the studies. A really good meta-analysis provides an understanding of how well the authors think they have done this, in relation to the quality of the studies included and the features of the studies they have explored.

Lack of Replication: A Serious Problem in Education

There is almost no replication in education research. A recent study suggested that it is as low as 0.13% (Makel and Plucker, 2014). The authors distinguish between direct and conceptual replication where the former repeats the experimental procedure and copies the initial study as closely as possible (design, sampling, methods) whereas the second uses different methods to test the same underlying hypothesis. They found 63 direct and 159 conceptual replications from 164,589 articles. Both approaches to replication are important to move a field forward systematically. It is problematic because we don't know how accurate research findings are and how much they might be expected to vary according to the particular conditions studied. Makel and Plucker's work indicates that 71% of direct replications and 66% of conceptual

[9] http://www.prisma-statement.org

ones were successful, suggesting that we should be sceptical about the results of as many as one in three empirical studies in education.

Why is there such a problem? I think you can look at this from two viewpoints. First, a number of issues disrupt rigour at the level of individual studies and researchers. The tyranny of statistical significance and publication bias, the belief that novelty is more important than accuracy, the nature of individual careers and academic success all create a research ecology which makes cumulative rigour difficult to achieve. The second viewpoint is a cultural one. Education is not sufficiently valued to merit this level of scrutiny and spending. Replication is expensive and potentially wasteful of resources, particularly from the perspective of funders. You feel like you are marking time, rather than moving forward. Lack of replication, however, is more inefficient overall, creating a field as a whole which moves forward very slowly, if at all. Gene Glass, who coined the term meta-analysis (see Chapter 2) talked about the 'fragility' of findings in education:

In education, the findings are fragile; they vary in confusing irregularity across contexts, classes of subjects, and countless other factors. Where ten studies might suffice to resolve a matter in biology, ten studies on computer assisted instruction or reading may fail to show the same pattern of results twice. This is particularly true of those questions that are more properly referred to as outcome evaluation than analytic research. (Glass, 1976: p. 3)

Meta-analysis is not replication. Meta-analyses aim to synthesise previous findings and pool them, with a focus on convergence (hopefully) and understanding variation. The aim of replication is to verify the accuracy of previous findings and to establish any underlying variation in effects. This is an important distinction. Ideally you would only want to meta-analyse findings which have been directly and successfully replicated. If up to one-third of findings really cannot be replicated they add substantial noise to a meta-analysis and increase the uncertainty around the findings. Meta-analysis is closer to conceptual replication than direct replication, but we need to be clear that it is not conceptual replication either. The authors of the studies were not testing the same hypotheses. We can look at the findings from a meta-analysis and see which features of the methods and of the intervention help to explain variation in a moderator analysis. This looks at the variation or heterogeneity between studies and tries to find associations which can account for or 'explain' this variation. But a meta-analysis is only as good as the included studies and can't effectively guard against systematic biases in a field. However, until we have systematic replication, both direct and conceptual, meta-analysis is the best option we have.

Chapter Summary

Meta-analysis can help us do a number of things. It can help make sense of diverse findings with an estimate of the overall average across different studies. Rather than relying on a single study, it combines the findings from similar studies, to try to answer a question more conclusively. It can also help to explain variation by identifying patterns in effects associated with various features in the underpinning research studies. Do longer or shorter interventions have larger effects? What about for younger or older learners? Is the impact the same or different when we look at different outcomes, such as single words or understanding sentences in reading or number fact recall or problem-solving in mathematics? Meta-analysis helps to build up a picture by supporting the synthesis of findings across studies. It can contribute to developing a more coherent picture of the research and evidence in a particular field. It indicates where there is already a body of evidence and can identify questions for further research. It can also help us understand more realistically what kind of improvements we can expect for learners. A highly successful intervention with an effect size of 0.8 is still only likely to help 28% of the children involved compared with what they would have experienced in the control condition.

Final Thoughts

I chose the introductory fable at the beginning of the chapter for two main reasons. First, because it highlights why we need meta-analysis to help us see past individual researchers' personal enthusiasms and interests. At times, we need to stand back to get an overview of the bigger picture in education research. What are the trends in research findings? What is the overall effect, on average? What, if anything, explains variation in the impact of a particular programme or approach? Meta-analysis gives us this view point and lets us hover over a number of studies to see the patterns and bigger picture that emerges when we aggregate or synthesise findings across studies. I also chose the fable 'Nothing Like Leather' because it acts as a reminder that the meta-analyst can be like the currier and claim 'there is nothing like meta-analysis'. Although a powerful technique for aggregating and synthesising research findings, it too is not a panacea. To understand its strengths and weaknesses, I think it is useful to understand the history behind its evolution. So, it is to the development of the approach and to some of the history of the underlying statistical techniques that I turn in Chapter 2.

2   A Brief History of Meta-analysis

Key questions
What are the origins of meta-analysis?
What can we learn from the evolution and development of
meta-analysis?

The Origins of Meta-analysis

As we saw in Chapter 1, it can be difficult to tell from the result of a single
piece of research whether or not something is a good idea. This is true in
medicine, with one example being the use of different beta-blocker drugs
where it took a number of studies to decide that such medication was
effective at reducing further heart attacks (Yusuf et al., 1985). This is even
more true in social science where it is never likely that one study, however
large or robust, will be definitive, due to the variation in contexts and
social settings. We therefore need to be able to identify any trends in
research findings by combining them to see what kind of patterns emerge.
This chapter traces the development of meta-analysis in education and
the history of meta-meta-analysis or 'meta-synthesis' in more detail,
where the temptation is not just to draw conclusions about similar studies
using a quantitative synthesis but to aggregate or compare findings across
meta-analyses to understand the relative benefits of different approaches
on educational outcomes.

Meta-analysis is a technique used in reviewing and summarising the
findings of different research studies and involves the statistical combination of their research findings. In this book 'meta-analysis' refers to
a quantitative synthesis which pools or aggregates findings, using agreed
statistical techniques (Borenstein et al., 2009). This is from a series of
studies which have been identified explicitly and rigorously, usually by
systematic review techniques. One aim of such a technique is to help in
drawing conclusions about whether an intervention or approach, on
balance, is effective or not. It also seeks to explain variation in research
findings by identifying any patterns or significant associations with
features of interventions associated with greater or smaller effects

25

('moderators'). So, in understanding whether phonics is an effective approach for early reading there are a number of meta-analyses which have looked at the results of other studies and concluded, on balance, that such approaches are effective, as discussed in the previous chapter (e.g., Ehri et al., 2001; Torgerson, Brooks and Hall, 2006; Jeynes, 2008). Each of these meta-analyses identifies an overall average or 'pooled' effect (an 'effect size' of 0.41, 0.27 and 0.30, respectively: this is the standardised mean difference expressed in standard deviation units). The precise estimate of effect depends on what makes up the average with the detail of the review and meta-analytic procedures and their estimates varying due to the detail of the different research questions and precise inclusion criteria. Each meta-analysis also draws slightly different conclusions about such things as the value of starting phonics at a younger age (Ehri et al., 2001) or the lack of evidence for synthetic over analytic approaches (Torgerson et al., 2006) or the robustness of the findings across studies of different quality (Jeynes, 2008), depending on the research questions and the way the different studies are compared and coded. Different aspects of implementation matter between studies (Durlak and DuPre, 2008). Although meta-analysis is not restricted to intervention research, this chapter focuses on studies with experimental designs with clear and deliberate comparisons between those who received and intervention of approach and those who did not. This kind of research is usually designed to answer efficacy or effectiveness questions such as does it work? Meta-analysis is now more widely used in medicine and psychology, but the term was first coined for educational research, and the underpinning statistical ideas date back over a further 70 years or so. This account takes an historical perspective and traces the development of meta-analysis in education and the evolution of 'meta-synthesis' or 'meta-meta-analysis' in more detail, where the temptation is not just to draw conclusions about similar studies using a quantitative synthesis but to aggregate and compare findings across meta-analyses to understand the relative benefits of different approaches on educational outcomes.

Meta-analysis combines or 'pools' estimates from a range of studies by averaging them and can therefore produce more widely applicable and generalisable inferences than would be possible from a single study. In addition, it can show whether the findings from similar studies vary more that would be predicted from their scale so that the causes of this variation (known technically as 'heterogeneity') can be investigated using further analysis to see what features are associated with specific effects, such as the length of time pupils studied, or the importance of training and support, or the use of particular resources, by drawing on data from across the included studies and looking for correlations. These factors are known as 'moderators' because the analysis identifies how much they influence or 'moderate' the overall effect. This is an important point, especially for education research where the results from small studies can be combined to provide answers to questions without being so dependent on the statistical significance of each of the individual studies, which is directly related to sample size and has other problematic assumptions (Nickerson, 2000). Many small studies with small or medium effects may not be likely to reach statistical significance. So, if you review the field by simply counting how many were statistically significant or by undertaking a narrative review, you may be easily misled into thinking that the evidence is less conclusive than if you combine these studies into a single meta-analysis to look at the overall pattern (see Cooper and Rosenthal (1980) and Hedges and Olkin (1980) for nice empirical demonstrations of this).

The Emergence of Meta-Analytic Approaches and Techniques

The history of the evolution of meta-analysis as a statistical technique to find more conclusive answers to research questions like this by combining findings is a fascinating one (Hunt, 1997; O'Rourke, 2007). It involves a number of academic characters, crosses disciplines and took nearly 60 years from conception in the early 1900s to its birth and naming in the 1970s. Once meta-analysis emerged as fully formed technique, its use expanded rapidly in a number of fields, particularly medicine and psychology. A number of issues in the development of the approach provide some salutary warnings, however, for contemporary use.

A British mathematician, Karl Pearson, appears to have been the first to think of ways to combine numerical data from different studies. He is considered the pioneer of mathematical statistics, and some of his techniques are still used today, such as the 'product-moment' correlation coefficient and the chi-squared test. In 1904, in the second volume of the *British Medical Journal*, he published an analysis of the incidence and mortality rates of typhoid fever among soldiers in the British Army in India and South Africa. Results were available for soldiers who had volunteered for inoculation and other soldiers who had not volunteered. He wanted to know whether combining findings from a series of small studies would help answer the question about the effectiveness of inoculation but was also curious about what was causing variation in findings, as vaccination sometimes appeared useful but sometimes did not (see also Susser, 1977). These two aims of combining findings for greater certainty but also finding out what accounts for any variation are the core concepts of meta-analysis.

Pearson presented the results of his work in a table where each study was represented by its own row showing the measure of effect, together with a measure of the within-study uncertainty (see Figure 2.1). The last row gives an average correlation as an average or pooled estimate of the effect. Again, this prefigures the development of the 'forest' plot which is now the classic way to present the findings of a meta-analysis (see Chapter 3). We can certainly recognise Pearson's table as 'meta-analysis' (Pearson, 1904: p. 1244), though without specifically being named as such, and it does not contain estimate of the uncertainty of this pooled effect (if we are being pedantic). This was not enough for Pearson however. He also wanted to understand what caused the variation or heterogeneity in the effectiveness of inoculation, so he looked for possible explanations, such as that soldiers who had volunteered for inoculation might have been at lower initial risk of developing the disease. He considered that this uncertainty

INOCULATION AGAINST ENTERIC FEVER:
Correlation between Immunity and Inoculation.

I. Hospital Staffs ...	...	...	+ 0.373	±	0.021
II. Ladysmith Garrison	...	...	+ 0.445	±	0.017
III. Methuen's Column	...	...	+ 0.191	±	0.026
IV. Single Regiments	...	...	+ 0.021	±	0.033
V. Army in India ...	...	...	+ 0.100	±	0.013
Mean value	...	...	+ 0.226		

Correlation between Mortality and Inoculation.

VI. Hospital Staffs ...	...	...	+ 0.307	±	0.128
VII. Ladysmith Garrison	...	...	— 0.010	±	0.081
VIII. Single Regiments	...	...	+ 0.300	±	0.093
IX Special Hospitals	...	...	+ 0.119	±	0.022
X. Various military Hospitals	...	...	+ 0.194	±	0.022
XI. Army in India ...	...	...	+ 0.248	±	0.050
Mean value	...	...	+ 0.193		

If we except IV and VII, the values of the correlations are at least twice (in the very sparse data of VI) and generally four, five, or more times their probable errors. From this standpoint we might say that they are all significant, but we are at once struck with the extreme irregularity and the lowness of the values reached. They are absolutely incomparable with the fairly steady and large values of the vaccination correlations obtained for different epidemics and towns. The effect of enteric inoculation is evidently largely influenced by difference of environment or of treatment.

Figure 2.1: Pearson's findings (Pearson, 1904: p. 1244)

might be answered by further analysis but also proposed 'an experimental inquiry' by inoculating every second volunteer as a randomised trial. Although it took another 60 years to develop more formally the techniques that he considered, Pearson set out a visionary approach for the use of randomised controlled trials and to aggregate findings across such trials (see also Fraser et al., 2007).

Another pioneer of statistics in scientific inquiry in the early part of the twentieth century was Ronald Fisher. He built on Pearson's work on correlation with the development of techniques like analysis of variance (ANOVA) in his work on agriculture at Rothamsted Experimental Station in Hertfordshire in England. In his textbook (Fisher, 1935), he gives an example of analysis of multiple studies, identifying the probable and real concern that fertiliser effects will vary by year and by location. Fisher's influence on the development meta-analysis was important. He laid the groundwork for the analysis of multiple studies in his final book, *Statistical Methods and Scientific Inference*, published in 1956. In this he encouraged researchers to summarise their findings clearly and rigorously, which would make the comparison and aggregation of cumulative estimates across studies easier, almost the same as if all of the data were available for reanalysis (Box et al., 1978).

A False Start?

One of the major challenges faced by researchers in the last century was the ever-increasing quantity of published research in almost all research fields. This needed new methods to synthesise and summarise the accumulating results. The first systematic attempt at this came in an unlikely area of psychology. In 1940, Joseph Gaither Pratt and Joseph Banks Rhine, based at Duke University in Durham, North Carolina, in the USA, published a book with some of their colleagues which reviewed 145 reports of extrasensory perception (ESP) experiments which had been published between 1882 and 1939 (Pratt et al., 1940). In chapter 4 they discuss and summarise 'the full range of available trials made to test the ESP hypothesis' from experimental studies. They took a critical look at the data from the point of view of the design and conditions of the experiments to address general issues of whether the findings could have happened by chance or by normal perception of sight, touch and hearing by those involved in the experiment. They also considered experimental errors and clustered results from similar experiments for subgroup analysis. For its time, it was a rigorous and clearly documented attempt to address the question of whether there was 'unequivocal evidence for the occurrence of ESP'. One of the approaches included in their review to demonstrate the robustness of the analysis was an estimate of the influence of unpublished papers on the overall pooled effect. Today this is often called the 'file-drawer' problem of publication bias on the basis that non-significant studies were thought to languish in a filing cabinet drawer, resulting in bias in the overall conclusion resulting from the omission of non-significant or negative findings. The effect of published studies, and lack of publication of what are seen as unsuccessful studies, remains a problem today. The work of Pratt and Rhine also sets a further challenge for advocates of meta-analysis. They concluded, on the basis of the evidence they summarised, that ESP *did* exist. Today we might criticise their analysis for a series of problems with the underlying studies; for some this was the design and conditions of the experiment, for others there was both publication bias and probably some selective reporting. The most important reason we are sceptical, however, is that, despite attempts, the findings have not been replicable. This suggests that the significant findings in the studies they accumulated had occurred by chance (as you would expect for one in 20 studies at the 95% level), poor design, error or even cheating. Ben Goldacre, in both *Bad Science* (2010) and *Bad Pharma* (2014), argues that some of these issues, particularly selective publication and publication bias, are still a problem in medical research today. Scepticism about the conclusions reached by Pratt and Rhine may well have influenced people's views about the soundness of the methods that they used, and this may have slowed the adoption of the systematic accumulation and analysis of evidence more widely.

The reason that this ESP 'meta-analysis' is so important is that it reminds us of three things. First, that the picture created by accumulating research findings over time may systematically present an incomplete or biased view. The reasons for this may be complex and come from a number of sources. Even a rigorous analysis, by the standards of the time, may not uncover this bias. The second challenge is the position of the researchers in relation to the evidence and argument. The rigour and transparency of Pratt and Rhine's analysis and the quality and the care with which they undertook their experiments makes it unlikely, in my view, that they were knowingly trying to deceive their audience. However, they were advocates of ESP and of parapsychology more widely and were trying to convince the scientific community that ESP was a genuine phenomenon. This may have influenced their decisions in the choices they made about their review, introducing researcher bias, at least at the unconscious level. The third issue is the importance of replication, particularly independent replication. This is particularly important in areas like education where the nature of an intervention is rather more difficult to specify and repeat than the contents of an inoculation or formulation of a tablet or even the design of a card-guessing experiment in ESP. Lack of replication remains a problem today (Open Science Collaboration, 2015).

There is an additional challenge in accumulating research evidence over time, which affects some aspects of medicine and some of the social processes and interactions such as those involved in education. Typhoid or enteric fever is usually caused through contamination of drinking water with a particular variety of *Salmonella* bacteria. Better sanitation has reduced the incidence of the disease. We understand considerably more now about the different strains of typhoid and the problem of asymptomatic carriers, like 'Typhoid Mary', who was first identified three years after Pearson's inoculation study was published. We also know through a systematic review and meta-analysis that today's vaccines are between 51% and 55% effective (Fraser et al., 2007; Anwar et al., 2014). Typhoid, however, is treated with antibiotics which influence the evolution of the bacterium, developing resistance and changing the response to vaccination. In education, we know little about how teaching practices evolve and the development of 'pedagogical resistance' to intervention, but we do know that the impact of interventions varies considerably according to context and have only just started to understand the factors that are associated with this variation. Through the 1950s and 1960s, a number of researchers tried parametric statistical techniques to summarise results from different studies which were reported as correlations or percentages (Kulik and Kulik, 1989). Underwood (1957) adopted percentages to summaries data on interference and lack of retention of information in studies of memory and forgetting. Erlenmeyer-Kimling and Jarvik (1963) applied statistical techniques to 99 correlation coefficients representing degree of similarity in intelligence of individuals who were related. Kulik and Kulik (1989) identified how these approaches to quantitative review are similar to a meta-analysis in three ways. First, a number of studies are summarised from the literature; second, they report findings from similar studies on a common scale, and finally they tried to explain some of the variation in study results according to specific features in the studies they reviewed.

To summarise this section, it is clear that by the beginning of the last quarter of the 20th century researchers were looking for ways to combine findings from similar studies to provide a more secure or more convincing answer to an overarching question. However, in doing this there were a number of challenges. First and foremost is the conceptual question about whether it makes sense to combine the studies. Or, to put it another way, what question can reasonably be answered by blending the results of different research enquiries? Is there likely to be a pattern of bias in the results? How systematically have these studies accumulated or how patchy is the evidence? Are there any gaps in this evidence which might lead to misleading conclusions? Are there any replications or independent evaluations where people have tried explicitly to see if the findings are repeatable?

The Birth of Meta-Analysis

Meta-analysis, in its current form, was not first undertaken in medical research. In 1976, in his presidential address to the American Educational Research Association, Gene Glass coined the term 'meta-analysis' to refer to 'the statistical analysis of a large collection of analysis results from individual studies for the purpose of integrating the findings'. With his colleague Mary Lee Smith, he had aggregated the findings from all of the psychotherapy outcome studies they could find in order to challenge the consensus, dominated by Hans Eysenck, that psychotherapy did not work. Eysenck was a professor of psychology at the Institute of Psychiatry, King's College, University of London and had previously reviewed the effects of psychotherapy by combining percentage scores and computing statistics from this data (Eysenck, 1952). Smith and Glass's (1977) analysis showed that 'the typical therapy trial raised the treatment group to a level about two-thirds of a standard deviation on average above untreated controls; the average person receiving therapy finished the experiment in a position that exceeded the 75th percentile in the control group on whatever outcome measure happened to be taken'. This represents an effect size of 0.6, leaving no real doubts about the value of psychotherapy.

Gene Glass credits Robert Rosenthal for developing the underlying metric of 'effect size' or the use of standard deviation units which are the measure used in most educational meta-analyses of impact studies that we now call 'meta-analysis'. In 1966, Rosenthal had published a book entitled *Experimenter Effects in Behavioral Research*, which contained calculations of a large number of standardised mean differences ('effect sizes') that he then compared across domains or conditions. Glass also acknowledges the impact of Benjamin Bloom on educational thinking more widely, with similarities between his approach in presenting aggregated graphs of correlation coefficients in his 1964 book *Stability and Change in Human Characteristics* though his contribution to meta-analysis is rarely noted. Indeed, his formulation of the search for teaching approaches as effective as one-to-one tuition was formulated in standard deviation units as 'the two-sigma problem' (Bloom, 1984). From the late

1970s through the 1980s the number of meta-analyses and methodological papers about the statistical techniques burgeoned with a consensus about the best approach emerging only slowly.

Of course, not all researchers accepted the approach: Eysenck, for example, called it 'mega-silliness' and expected it to be a short-lived distraction. Others criticised it because of the comparability issue claiming it combined 'apples and oranges' (see Slavin, 1986 for a discussion of this). Glass's (2000) riposte is telling: 'Of course it mixes apples and oranges; in the study of fruit nothing else is sensible; comparing apples and oranges is the only endeavor worthy of true scientists; comparing apples to apples is trivial.' This gets to the heart of the issue. If your answer is about apples and oranges, you need to know this and not believe it is just about apples, or oranges, or even all kinds of fruit. Any inference is directly related to what the meta-analysis contains. Understanding this is not always straightforward as it seems. In education, we use a large number of general terms. Take 'homework', for example. We are all confident we know what the word means, so identifying studies of homework should not be problematic. However, if you want to know if homework helps children to learn better you have to decide what counts as homework. Is a class of five-year-olds taking books home to practice reading with their parents a form of homework? What about learning spellings at home? How about 'homework clubs' where children do their 'homework' at school, *before* they go home? What about reading in preparation for a lesson, what might now be called flipped learning? Or learning multiplication facts over the weekend in preparation for a test on Monday morning? What about completing coursework for an examination at home? Are these really all the same thing? They are 'homework' only in the sense that they are not 'classwork' with a teacher present. Suppose you combine all of these types of studies and conclude that children who are assigned homework compared with similar children who are not given any do better, on average, in tests of their learning, what does this tell you? It does not mean that homework is always effective. It indicates that when people have experimented to find out if being given (and doing) homework helps, the broad answer is yes. It does not tell you that it will work in every instance in the future. If you know what kind of homework studies have been included, you will have a reasonably clear idea about what might be likely to work in your school or for your own children. It would also be helpful to know if there are any other implications in the research. Do reading studies show greater effects? What about the age of the children? What about frequency and regularity of homework set? A general answer is useful but is then is only the starting point for further questions and investigation.

This combination of studies is also a problem in another way which Eysenck (1978: p. 515) and others (such as Slavin, 1986) also took exception to. Eysenck was concerned at the lack of quality control in what he termed the problem of 'garbage in – garbage out' (p. 517). He thought that only studies of high quality should be included as less rigorous studies might introduce bias into the results. The challenge here is in deciding what counts as a high-quality study. Do you only include those with a rigorous design, such those with a randomisation which can demonstrate that they have effectively controlled for selection bias and reduced the risk of possible confounding variables and where the sample size meets a specific threshold and where the quality of the research process is adequately documented? Or do you include data from a wider range of experimental comparisons and check to what extent features of the research design and the quality of the reporting explain variation in the pooled effect?[1] This is not an easy decision. If you have sufficient studies you may be able to afford to reduce the quantity to assure the quality. However, you also have to be careful that this does not introduce other kinds of bias. Research and reporting quality are not necessarily related to the actual effectiveness of an intervention. A researcher may have had a great idea about how to improve reading but not been very good at evaluating and describing this. If the findings are consistent with more rigorous studies, additional data on what causes variation may usefully add to moderator analyses, which are often underpowered (Valentine et al., 2010). Discarding the data without investigating or establishing this seems premature, though it is important to have clear criteria for inclusion and to have a specification or protocol in advance for checking the relationship between study features and quality, particularly in a more inclusive study, so as to prevent chance findings being identified as significant, sometimes known as 'data dredging' (see, for example, Moher et al., 2009).

The Application and Uptake of Meta-analysis in Medicine

The development of meta-analysis as an essential technique to summarise and synthesise research findings across medical studies began a few years after Glass first coined the term in the 1970s. Particularly important here

---

[1] In medicine there is a reporting standard which is expected for randomised controlled trials, the Consolidated Standards for Reporting Trials, or CONSORT agreement.

was an innovative randomised trial conducted by Peter Elwood, Archie Cochrane and others (Elwood et al., 1974) to find out whether taking aspirin lowered the risk of further heart attacks and reduced the mortality rates associated with these (O'Rourke, 2007). The overall results suggested there was a benefit but were not conclusive. So, over the next few years, Elwood and Cochrane collected and combined results using meta-analysis as additional findings from new trials were reported. This aggregation left little room for doubt that taking aspirin after a heart attack was beneficial, and the findings were presented in 1980 in an anonymous *Lancet* editorial, but penned by the British medical statistician Richard Peto.

Peto and his colleagues used a further example with data from randomised trials of beta-blockers following heart attack to encourage medical practitioners to consider aggregation of data from randomised trials systematically and to combine quantitative estimates of the effects of comparable medical treatments. These developments started a debate, similar to one developing in the social sciences, about the best way mathematically to 'average' or estimate the aggregated findings in a single figure estimating the overall impact of different interventions or treatments, as an effect size. Peto argued for estimating a weighted average of the different effects when the effects were not identical, so treating the meta-analysis as if it was a larger single study: the 'fixed' effect model (based on inverse variance where studies with smaller variance (standard error) contribute more than studies with larger variance). A more conservative approach is to think of each study being a slightly different version of an intervention, with its own random variation which needs to be accounted for and where both the variation *within* studies and *between* studies is taken into account). I think of this as hedging your bets between the studies and being a bit more conservative about your conclusions because you don't know what causes this variation. This kind of 'random effects' model was advocated by meta-analysis pioneers like Larry Hedges (1983) and was developed and promoted to medical researchers by Rebecca DerSimonian and Nan Laird, who also provided simple approximate formulas for Cochrane's formal random effects model (DerSimonian and Laird, 1986). For a presentation of the arguments for fixed effect and random effects models and methods of calculation, see Borenstein et al. (2009).

Similar to developments in social sciences a few years earlier, these developments in medical research led to an explosion of research and publication presenting empirical findings, developing the methods and promoting the work to practitioners (O'Rourke, 2007). One key difference in clinical work was the focus on quality assessment compared with work in social sciences. As the technical literature on meta-analysis expanded, the importance of being confident about the data included in a meta-analysis received increasing attention. So, using systematic approaches to identify and collect relevant information so as to reduce bias in reviews became more and more important. The greater statistical precision of findings from meta-analysis are not of much use if they are accurate to three decimal places but misleading! Even today this problem has not been solved in medicine.

The range of terminology used in medical reviews was confusing and towards the end of the last century, Chalmers and Altman (1995) argued that the term 'meta-analysis' should be applied only to quantitative synthesis, as adopted in this book. One of the main reasons for the rapid growth of meta-analysis is that it tackles one of the key challenges in reviewing research in that it can cope with a large number of studies which can overwhelm other approaches (Chan and Arvey, 2012). In addition, the statistical techniques to undertake meta-analysis form a set of transparent and replicable rules which are open to scrutiny and which have been accepted across a number of disciplines (Aguinis et al., 2011). One of the other drivers in medicine is that the conceptual issues in identifying the 'intervention' were arguably more straightforward. When considering the effects of aspirin or beta-blockers on heart function the definitions are clear and precise. Formulation and dosage can be described chemically. When thinking about the effects of phonics interventions on reading outcomes the challenge of precise definitions becomes clearer. My own view here is that this helped the field solve some of the technical and statistical issues, whilst in education we have struggled, often with good reasons, with definitions and concepts about what was involved in an 'intervention'.

The ability to include the wealth of studies available is particularly important when trying to draw cumulative inferences in a specific area of education research. The number of studies available to review in any area of education can be extensive, so techniques to aggregate and build up understanding of a field in terms of the impact of different interventions or approaches and what might explain variation so as to propose further research and test theories and hypotheses are invaluable. In fields like psychology and medicine meta-analysis is relatively uncontroversial as a synthesis method, with nearly 40 years of development of the various principles and methods involved, despite its initial origins in education.

Limitations and Challenges for Meta-analysis

There are limitations, of course, and perhaps the most important is the assumption that the data from evaluations are equivalent or at least comparable across studies. Here the key issue in education is a conceptual one (Lipsey and Wilson, 2000). Are the studies which are being compared actually the same in terms of the way that they have defined or implemented a particular approach? This also relates to the nature of the question being addressed. Asking whether phonics interventions are effective for beginning readers to catch up with their peers is different from asking whether phonics approaches are the best approach for beginning readers (when compared with other approaches to teaching reading). Some studies would be included in both reviews, but in one it may be helpful to combine studies in different categories (phonics, whole word, comprehension-led, whole language, etc.), and clarity about definitions and outcomes (such as decoding words or comprehending sentences) would be essential.

Another limitation is the so-called 'file-drawer' problem where studies with null or negative effects are not reported or are less likely to be reported because they are not thought to be interesting or to add anything new to a field. If a set of findings from research is systematically missing, in this case the null or negative studies, then meta-analysis will provide an inflated estimate of the overall effect. Most researchers in education work in fields they are enthusiastic about, they are like the currier in the poem which introduced Chapter 1. They may well be less likely to report or write up studies which do not show positive effects. This may not be deliberate deceit, rather people may believe they know why a study was unsuccessful and seek to remedy what they saw as a problem in the design or implementation. Another feature of publication bias results from the influence of statistical significance on publication. Smaller studies need larger effects to meet the criterion of statistical significance. If there is bias which suppresses the publication of non-significant findings and there is a predominance of small-scale studies (both of which are evident in education research), then the large effects from small studies which are statistically significant will dominate in a meta-analysis. The result will be an overestimate of the effects. Additionally, we have to be cautious with many evaluations of impact in education where the nested or clustered nature of schooling is not taken into account (Raudenbush, 1997; Campbell et al., 2012). Pupils work in classes which are in schools and both the class they are in the school they attend may influence the impact of different approaches. Analysis needs to take this into account or the effects may be overestimated and the apparent precision of the estimates may be overconfident (Xiao et al., 2016).

However, there are procedures to guard against potential biases through transparent and conceptually clear inclusion and exclusion criteria, careful searching and systematic review, consideration of heterogeneity of effects and examination for publication bias to understand the nature of the data included in a meta-analysis. This needs to be used to inform the interpretation of the findings. Although there are limitations to the application of quantitative synthesis as described above, the data from meta-analysis offer the best source of information to address cumulative questions about effects in different areas of educational research, as well as in understanding what might explain differences in effects. For these questions, the technique is relatively uncontroversial.

There are also technical issues involved. How comparable are results from different tests? Or test results from pupils of different ages? What about the methods used by different researcher? How comparable are these? Some of these will be considered in the next chapter in terms of understanding the limits of the use of effect size as a measure. As a rule of thumb, the more similar the ingredients (such as the intervention or approach, the tests used, the pupils and their characteristics, how long the research lasted, the research design and comparisons made) the more precisely the combined effects can be estimated and the clearer the claims can be made for the findings. The more varied the ingredients, the more cautious we should be about the precision of any estimate of effect and about what claims can be made about what the findings mean.

The random effects model, referred to in the previous section is theoretically more appropriate and statistically more robust for combining or averaging studies with pedagogical variation. This weights studies using both the within- and between-study variance to get a more representative average. I understand why this should be preferred in education research. However, this technique may also inflate the overall estimate in 'hedging its bets' between the studies, with the result that smaller studies will tend to carry more weight. If there is a bias towards smaller studies reporting larger effects, as appears to be the case in education, then this will be reflected in the pooled average effect size. With enough data, we could adjust for this (and other systematic variation) in educational meta-analyses.

In this short historical overview, it is not possible to include a number of important researchers who have developed statistical theory and practice underpinning meta-analysis. These include people like Thomas C. Chalmers, Jacob Cohen, Harris Cooper, Larry V. Hedges, John E. Hunter, Ingram Olkin, Nambury S. Raju, Robert Rosenthal and

Frank L. Schmidt to name but a few. Without their contributions, the field would not have developed the way it has.

Chapter Summary

The history of the evolution and development of meta-analysis is instructive. It reminds us of both the possible benefits and potential pitfalls involved. Aggregating results to get the bigger picture is the main aim of meta-analysis. However, this aggregation has risks. The most important issue here for education is to have conceptual clarity and precise definitions so that we can be sure about what is being compared and what claims can be made about the findings. It is also important to understand the risks of bias in the studies included. A meta-analysis is just a way of averaging the effects from different studies, albeit according to some technical rules. Any average depends on how good the individual estimates are. Bias is an issue in any form of averaging. In meta-analysis, there are some particular risks from bias, such as publication bias, which can to some extent be estimated and controlled for. There are also other potential biases, such as those underlying a particular field, as we saw with ESP, which may be harder to discern without the hindsight of history.

Final Thoughts

The history and evolution of any scientific idea is often instructive. Statistical techniques were developed to solve particular contemporary problems, and those who developed them were often well aware of the limitations and compromises involved in the techniques. Once the techniques become accepted, we perhaps forget the compromises involved. I find it helpful in understanding the strengths and weaknesses of meta-analysis to know more about the development of trials in medicine and the pioneering work done in this field.[2] This is not because I think education is, or should be, like medicine. A classroom is much more complex than a consulting room. I do think that a more rigorous approach to identifying cause and effect will help tackle some of our educational challenges and make education less susceptible to political

---

[2] The James Lind Library has been developed to illustrate the evolution of fair tests of treatments in health care (http://www.jameslindlibrary.org). The website is an invaluable source of information about the history of experimental trials. My own personal favourite account of the history of meta-analysis is Morton Hunt's 1997 book *How Science Takes Stock: The Story of Meta-Analysis*, which provides an accessible account of its development.

whim and personal preferences. It will never be deterministic, but we can look at patterns of effects over time to develop our understanding of what is likely to be of benefit to inform professional decisions about children's and young people's learning. More importantly perhaps, this would help us to understand what is not likely to be useful. Knowing the odds of an approach being helpful for different ages, subjects and outcomes in different schools and for different teachers strikes me as a worthwhile goal for research synthesis in education. The next chapter picks up these ideas and explores the value of not only using meta-analysis, but exploring what we can learn by comparing meta-analyses.

3 Meta-synthesis in Education
What Can We Compare?

Key questions
What is reasonable to compare?
What are the patterns in the findings from meta-analysis?
How much difference do interventions make?

'It's like comparing apples and oranges; they're both delicious.'
Cyd Charisse (1922–2008)

Introduction

This chapter will look at the use of meta-analysis to compare findings
from less similar studies to draw more general conclusions about what is
effective in education. I'll discuss some of the issues in interpreting
the relative difference between studies or between meta-analyses in
educational research. Summaries of key meta-syntheses (or meta-meta-
analyses) will be presented to identify some of the purposes of the
approach and the emerging patterns resulting from synthesising the findings from meta-analyses. It will present an overview of the challenges
inherent in the approach to understand what can reasonably be inferred
from the patterns of effects from intervention research. This is to put into
context the relative benefits of different approaches as summarised by
comparative meta-analysis, and to provide a picture or map of where
there is evidence about intervention effects for teaching and learning in
schools.

Meta-analysis and Meta-synthesis

It is tempting to look at results across different kinds of studies with
a common population in order to provide more general or comparative
inferences. A comparative meta-analysis, in this sense, compares effects
between different kinds of intervention studies and between meta-
analyses. It aims to answer the question 'Does X work better than Y?'
rather than the more specific 'Does X work?'. In this sense, a comparative
meta-analysis is a single meta-analysis where more than one intervention

41

or approach is included to identify which is more effective. A comparative meta-synthesis is defined as an overarching analysis where inferences are drawn by comparing findings across meta-analyses. This comparative approach is, of course, vulnerable to the classic 'apples and oranges' criticism, which argues that you can't really make a sensible comparison between different kinds of things. The key point here, as noted above, is that any inferences that you can make are at the level of the aggregation of the synthesis. In studying apples and oranges together you can consider what they tell you about fruit, similar characteristics (such as seeds for reproduction, developing from the female parts of the flower, protecting the seeds, developing an edible coating to aid dispersal) and variation (relative size, the nature of the fleshy covering, the detail of the different seeds and the like), but not conclusions specific to oranges (such as segmentation of the fruit, juiciness of the flesh, oily skin and the like). Similarly, in combining an analysis of different approaches to improving reading you can draw inferences about the effectiveness of, say, reciprocal teaching, compared with inference training (see, for example, Pearson and Dole, 1987) or comparing conceptual change approaches between science and reading interventions (Guzetti et al., 1993), assuming comparable populations and sufficiently similar interventions and designs. However, you can't draw conclusions about features which are not designed to be similar such as the impact on specific groups of pupils if the studies contain a wide range of different samples of students. Another example is the teaching of writing for primary-age pupils (Graham et al., 2012) where the meta-analysis indicates that teaching strategies, adding self-regulation to strategy instruction, teaching text structure, the use of creativity/imagery instruction and the teaching transcription skills are all important features of the effective improvement of writing but not the specific components of such strategies or which aspects of creativity and use of imagery were beneficial.

A number of researchers have attempted to take meta-analysis a step further than this by synthesising the results from a number of existing meta-analyses and producing what has been called a 'meta-meta-analysis' (Kazrin et al., 1979), a 'mega-analysis' (Smith, 1982), 'super-analysis' (Dillon, 1982), 'super-synthesis' or 'meta-synthesis' (e.g., Sipe and Curlette, 1996) to draw comparative inferences. This remains controversial in educational research, and there is a clear separation of types within these studies. Some use the meta-analyses as the unit of analysis in order to say something about the process of conducting a meta-analysis and identifying statistical commonalities which may be of importance (e.g., Ioannidis and Trikalinos, 2007; Lipsey and Wilson, 2000). Others, however, attempt to combine different meta-analyses into a single message about a more general topic than each individual meta-analysis can achieve (e.g., Walberg, 1982; Bloom, 1984; Hattie, 1992; Sipe and Curlette, 1996). Even here, there appears to be a qualitative difference, and some retain a clear focus, either by using meta-analyses as the source for identifying original studies with an overarching theoretical perspective (e.g., Marzano, 1998) in effect producing something that might best be considered as a series of larger meta-analyses rather than a meta-synthesis. Others, though, make claims about broad and quite distinct educational areas by directly combining results from identified meta-analyses (e.g., Fraser et al., 1987; Hattie, 1992; Sipe and Curlette, 1996). In terms of the apples and oranges analogy, this is a little like asking which fruit is best for you, as a lot depends on what you mean by 'best' and how this is measured. You might need to look at a range of studies and comparisons to find out.

In the following section a number of these 'meta-meta-analyses' or 'meta-syntheses' are reviewed to identify some key characteristics and limitations. The first was published just over ten years after Gene Glass set out a methodology to aggregate findings across studies, when a team involving Barry Fraser, Herbert Walberg, Wayne Welch and John Hattie undertook an extensive synthesis of evidence in which they summarised the findings from 226 meta-analyses (Fraser et al., 1987), incidentally indicating the rapid uptake of meta-analysis as a new technique. The main purpose of Fraser and colleagues' synthesis was to report the findings as an empirical test of Walberg's own educational productivity model, based mainly on quantitative syntheses of various aspects of previous research. Walberg's (1980) model emulated an economics model of productivity and proposed to explain students' academic achievement on standardised tests as a function of student ability and motivation, instructional quantity and quality, home and classroom environments, and age. The validation work was published as a 100-page monograph in the *International Journal of Education Research*. A chapter presents a research synthesis of several thousand individual studies to identify aptitudinal, instructional and environmental variables which consistently and extensively influence student learning. Another chapter focuses on meta-analyses of individual studies in science teaching and learning to identify the effects of contextual and transactional influences on science learning. Then the various salient features of contemporary models of student learning were used to structure a synthesis of 134 meta-analyses of achievement outcomes and 92 meta-analyses of attitude outcomes. The results of this impressive synthesis of meta-analyses was then used to identify to what extent the empirical evidence supported Walberg's nine-factor model of educational productivity.

John Hattie has been a pioneer in this field and in 1992 took this work a step further by summarising the 134 meta-analyses which had been identified and reported earlier in his collaboration with Fraser, Walberg and Welch. The synthesis consisted of 22,155 effect sizes computed from 7,827 primary studies, which represented between 5 and 15 million students. His aim was to show how findings from more than 30 years of educational research indicated 'the effects of innovation and schooling' but could also be used to 'provide insights for future innovation', as well as to resolve 'contrary claims about the effectiveness of schooling, by demonstrating how different points of comparison are used by each group' (p. 5).

As a key contribution to the field he introduced a 'universal continuum' (p. 6) in this analysis as a basis for this assessment, with a scale expressed in standard deviation units and with results from the meta-synthesis placed on this scale. Although the effect sizes were reported as correlations in the paper with Fraser and Walberg, Hattie converted these into standardised mean differences, as Bloom had previously done for tutoring in 1984. The average effect size across the meta-analyses was 0.40 (with a standard deviation of 0.13). The largest effect sizes were for those interventions providing feedback. The effect sizes for reinforcement, for remediation and feedback, and for mastery learning were 1.13, 0.65 and 0.50, respectively. The lowest effect sizes involved individualisation (with an average effect size of 0.14), whilst programmed instruction, another approach to personalisation, yielded an average effect size of 0.18. Hattie identified three overall findings. First, innovation, as the deliberate attempt to improve the quality of learning, can be related to improved achievement and is an underlying theme in the majority of these effects. Second, feedback is the most powerful single influence which improves achievement. Third, the least successful innovations were those that tried to individualise instruction. He saw this as important because, at the time, pupils spent about two-thirds their time in school working on their own. In terms of the overall messages, Hattie was concerned that the underlying studies were of variable quality as well as using different outcome measures. He also indicated that the synthesis did not suggest that effects on achievement are necessarily cumulative. His conclusion was about the value of the comparative information from such an approach and that the 'continuum highlights the importance of assessing competing theories – that is, to compare various innovations' (Hattie, 1992: p. 11).

Four years later, Sipe and Curlette (1996) published a review which explicitly aimed to develop the methodology of summarisation and meta-synthesis in terms of the evidence related to educational achievement. They undertook a systematic search for meta-analyses, applied rigorous inclusion criteria (excluding 324 of the 427 meta-analyses they had identified) and then described the background, methodological and contextual characteristics of the 103 studies they included, involving over 4,000 primary research studies. They also undertook an explanatory analysis, similar to Fraser and colleagues (1987), in relation to Walberg's (1982) educational productivity theory. They estimated that fewer than 10% of the meta-analyses in their meta-synthesis overlapped with those in Fraser et al. (1987) and Hattie's (1992) review, because the earlier reviews had included meta-analyses from 1976 to 1985 whilst theirs included meta-analyses published between 1984 and 1993 (p. 648). (For more details of other early meta-syntheses and a critique of 18 such studies see Sipe and Curlette (1996: pp. 597–612).)

Their overall conclusions about student achievement were very similar to those reported in Hattie (1992). The five highest effect sizes were for vocabulary instruction (ES = 1.15), accelerated instruction (ES = 0.88), mastery learning (ES = 0.82), direct instruction (ES = 0.82), and note-taking (ES = 0.71). One other meta-analysis ranked in the five highest effect sizes related to pupils' motivation and self-efficacy beliefs (ES = 0.82). In terms of what was unsuccessful, the five curriculum interventions with the lowest effect sizes were ability grouping (ES = −0.04), the Frostig Program (ES = 0.02) (this was an approach developed in the 1960s by a pioneer in the field of learning difficulties, Marianne Frostig, which aimed to develop visual perception so as to improve reading), matching teachers and students cognitive styles (ES = 0.03), factual adjunct questions (ES = 0.08) (these are questions added to an instructional text to influence what is learned from the text), the Intermediate Science Curriculum Study (ES = 0.09) (this was a science curriculum developed by Florida State University) and 'whole language' approaches for literacy (ES = 0.09) (see Chapters 1 and 5, Table 5.5). These programmes produced effect sizes near zero, which can be interpreted as essentially no effect of the interventions on the treatment groups. They also observed that, similar to the findings of other reviews of meta-analyses (Fraser et al., 1987; Hattie, 1992; Kulik and Kulik, 1989; Lipsey and Wilson, 2000; Wang et al., 1993). Large positive or negative effects were relatively rare among curriculum interventions. Over half produced small effects whilst one-quarter produced moderate effects, with Sipe and Curlette commenting 'apparently most educational treatments work, but their effects are modest' (p. 653). Importantly, they explored methodological features in their database and noted lack of consistency in the estimates of effect, indicating considerable variability within the categories they used. They thought that this variation (what we would now describe technically as unexplained heterogeneity) might result from differences in implementation of the treatment, design of the study or population included in the study. They suggested that the spread of effects was important to consider as it represented a measure of the stability of the effect. They noted (p. 654) that the 'continuum of effect sizes gives educators the opportunity to view the effects of many different educational programmes and compare between and among types of programmes' and that (p. 647) 'no other methodology provides the ability to examine the statistical results of so many research studies at one time'. These observations remain true 20 years later.

In 1998, Robert Marzano published 'A Theory-Based Meta-analysis of Research on Instruction' in a report for the US government-funded Mid-Continent Regional Educational Laboratory (McREL). This synthesised research findings from more than 100 meta-analyses and other studies involving over 4,000 comparisons of experimental and control groups (Marzano, 1998). One of the main goals of the synthesis was to identify instructional strategies which should have the greatest likelihood of enhancing achievement for all pupils, across subject areas and age groups. The analysis revealed that there was considerable variability across the studies in terms of how the instructional strategies were defined and how their use in the classroom was described. Marzano is critical of what he describes as the 'brand name' approach in meta-synthesis where very broad categories of educational approaches represent the popular labels which are sometimes given to quite complex interventions with a range of salient features or 'active ingredients'. As an example, he cites a meta-analysis conducted by Athappilly, Smidchens and Kofel (1983) and included in Fraser et al. (1987) where one 'brand name' used is 'modern math' but where a number of different components contribute to an aggregated effect size. Features in 'modern math' such as the 'use of manipulatives', which had an effect size of 0.51, could be distinguished from 'direct instruction in concepts and principles', which had an effect size of 0.35 and was different from 'use of an inquiry approach', which had an effect size of only 0.04. Marzano argues that aggregating these into a single 'brand' masks potentially important findings about instructional effectiveness and that more-discrete categories are needed which are specific enough to provide guidance for teachers in terms of classroom practice. He describes the meta-synthesis as 'theory driven' in using four high-level categories for learning based on *knowledge*, learning which involves the *cognitive* system or the *metacognitive* system, and the *self-system*. Overall, he draws conclusions at the level of specific instructional practices, but also at this more abstract or theoretical level. What makes Marzano's approach important is the overall coherence of the theoretical framework, whilst also striving to be practical at the level of applicability in the classroom. Marzano's (1998) meta-analysis extends previous efforts to synthesise evidence from meta-analysis to provide an 'emerging picture of effective teaching and the effective teacher' (p. 135). He was disappointed that he did not achieve his original goal to provide a comprehensive review of the research on instruction. Even though over 4,000 effect sizes were included, he estimated that at least three times this number would be needed to provide stable estimates of the effect sizes for each of the instructional techniques and approaches in relation to particular educational goals. However, he was clear that the effective teacher is one who has clear instructional goals which are communicated both to students and to parents. He argued that instructional goals should not only include knowledge and skills but also aspects of elements of the cognitive, metacognitive and self-system that he had used in the overarching framework.

Marzano (1998: pp. 134–135) identified nine features of effective teaching which were consistent across the different kinds of educational goals reviewed in the meta-analysis (see Table 3.1):

Table 3.1: *Marzano's nine features of effective teaching*

1.	When presenting new knowledge or processes to students, provide them with advanced ways of thinking about the new knowledge or processes prior to presenting them.
2.	When presenting students with new knowledge or processes, help them identify what they already know about the topic.
3.	When students have been presented with new knowledge or processes, have them compare and contrast it with other knowledge and processes.
4.	Help students represent new knowledge and processes in nonlinguistic ways as well as linguistic ways.
5.	Have students utilise what they have learned by engaging them in tasks that involve experimental enquiry, problem-solving, and (presumably) decision-making and investigation.
6.	Provide students with explicit instructional goals and give them explicit and precise feedback relative to how well those goals were met.
7.	When students have met an instructional goal, praise and reward their accomplishments.
8.	Have students identify their own instructional goals, develop strategies to obtain their goals and monitor their own progress and thinking relative to those goals.
9.	When presenting new knowledge or processes, help students analyse the beliefs they have that will enhance or inhibit their chances of learning the new knowledge or processes.

Marzano's work is impressive for its overarching vision, both in terms of the coherence of the theoretical underpinning and in the level of detail he believed that it was possible to achieve with a systematic approach.

In what can be seen as the culmination of his work to date, Hattie (2009) synthesised more than 800 meta-analyses, including those in his 1992 study, and came up with some interesting further findings in his book *Visible Learning* (see Table 3.2). As before, he concluded that most things in education 'work' as the average effect size is about 0.4 (a mean effect which has remained stable since 1992). He then uses this to provide a benchmark for what works above this 'hinge' point. The fact that this figure has remained stable when aggregating quantitative data in education is interesting, but as well as representing the typical difference of bringing about change in education as Hattie argues, it may also be taken to show that differences of just less than half a standard deviation, on average, are of educational interest and worth investigating. It is less clear that the same things have the same effect over time. Older studies of peer tutoring, for example, tend to have larger effect sizes, but whether this is the result of lower evaluation quality, publication bias, allocation bias, researcher bias or genuinely reflects a change in the counterfactual conditions so larger effects are harder to achieve (see Lemons et al. (2014) for a discussion of the problem of the changing counterfactual). Hattie updated his 2009 list of 138 effects to 150 effects in *Visible Learning for Teachers* (2011), and then to a list of 195 effects for higher education (Hattie, 2015). His research is now based on nearly 1,200 meta-analyses, but the pattern of effects underlying the data has changed little over time even though some effect sizes were updated and we have some new entries at the top, at the middle and at the end of the list.

Table 3.2: *Hattie's top ten effects in Visible Learning*

Self-reported grades	1.44
Piagetian programmes	1.28
Providing formative evaluation	0.90
Micro teaching	0.88
Acceleration	0.88
Classroom behavioural approaches	0.80
Interventions for learning disabled	0.77
Teacher clarity	0.75
Reciprocal teaching	0.74
Feedback	0.73

Hattie's impressive scholarship is extremely valuable in putting findings together to provide a large-scale landscape of all the quantitative findings from educational attainment amenable to meta-analysis; however, it is not without its critics (Snook et al., 2009; Terhart, 2011; Higgins and Simpson, 2011; Simpson, 2017). The key assumption is that the research represented in the meta-synthesis is sufficiently evenly distributed by type, population and outcome that any differences which emerge represent differences in the educational themes, rather than differences in the research methods, measurements and target populations. As noted above, combining effect sizes of different kinds is risky. Intervention effects (improvement relative to a comparison or control group) should be distinguished from maturational differences for the students (single group designs). The design of the former takes growth into account between the pre- and post-test, the latter *only* accounts for growth over time. Correlational effects, such as the relationship between, say, homework and school performance (or, to be precise, looking at pupils who do different amounts of homework and comparing this with how well they do on tests of attainment) are rather different from homework intervention effects (where the impact of homework for some is compared with no homework for others). Looking at the relationship (correlation) between teacher clarity and pupil outcomes in an observational study is very different from measuring the impact of an intervention which aims to improve teacher clarity and assesses this by comparing intervention and control groups in a randomised trial. Both can be converted to the same metric, a standardised mean difference, but the underlying distributions of educational attainment in these studies are likely to be of different kinds so that, unlike comparing fruit, it may be more like comparing an apple with a carrot, or even a chair (Higgins and Simpson, 2011). In an observational study, we are less certain about the causal mechanism. It may be that higher performing classes of pupils allow the teacher to give clearer explanations and instructions. Improving teacher clarity may not bring about the same impact. Effect size estimates in terms of standardised mean differences depend on the distribution of the individual scores, as this forms the basis for the comparison, or the standard deviation 'units' in which effect size is measured.

Overall, we can see a trend in educational research towards the end of the 20th century and into the 21st which drives the need for aggregation of research findings, partly due to the ever-increasing numbers of studies for review, but also partly to identify more-secure messages through aggregation and synthesis. Once this becomes possible quantitatively through meta-analysis, a further level of comparative meta-analysis, or comparative meta-synthesis also becomes possible. The temptation to draw conclusions between findings from different approaches is strong, as we lack a consistent and coherent set of comparative studies. Education researchers became interested in this approach shortly after meta-analysis was accepted as a synthesis method. The appeal is even stronger for practitioners, and the comparative or relative benefits are even more important in terms of making informed decisions in the classroom. It is all very well knowing if something 'works', but teachers need to know how well it works and how well it works compared with other similar approaches. Comparative meta-analysis, or meta-synthesis, can address these issues of the relative benefit of different approaches, though a number of challenges remain to be solved to improve the accuracy and predictive value of the findings.

Advantages and Limitations of Meta-analysis and Meta-meta-analysis

All of the limitations of meta-analysis apply to meta-synthesis. Explicit inclusion criteria and a systematic search are essential so that what is included in any aggregation is clear. Whether you choose to be inclusive in identifying studies and then checking how much difference methodological rigour makes so as to maximise possible data for analysis, or whether you set a high-quality threshold for inclusion to ensure rigour and internal validity in the component studies is perhaps less important, though transparency is essential. 'Garbage in' is still likely to result in 'garbage out' as Eysenck noted. The classic apples and oranges problem is also particularly relevant, but again the answer is similar to the one about comparing fruit. Provided we remember that any findings or 'answer' applies at the level of the aggregation of the meta-synthesis, then the approach and any inferences may be warranted. So, for example, if we ask whether some approaches to improving children's comprehension are more effective than others, then we can compare the findings from different meta-analyses, assuming they have both included similar studies (such as on typically developing or non-typically developing populations of pupils, with similar sample sizes, using similar outcome measures). If we want to know whether the impact of feedback remains higher than individualised approaches to learning in recent studies, we could look at the findings from different meta-analyses and evaluate whether the inclusion criteria for each have produced a sufficiently comparable set of studies to warrant a clear conclusion.

Although there are limitations to the application of comparative quantitative synthesis (both comparative meta-analysis and comparative meta-synthesis as defined above) in this way, the data from meta-analysis offer the best source of information to try to answer comparative questions between different areas of educational research. There is very little replication in education research, so meta-analysis is an essential tool. It is also hard to compare different educational approaches without some kind of common benchmark. If you have two narrative reviews, one arguing that, say, parental involvement works and another arguing that digital technology is effective, and both cite studies with statistically significant findings showing they each improve reading comprehension, it is hard to choose between them in terms of which is likely to offer the most benefit. Meta-analysis certainly helps to identify which researched approaches have, on average, made the most difference, in terms of effect size, on tested achievement of students in a measure, say, of reading comprehension or another area of educational achievement. This comparative information should be treated cautiously, but taken seriously. If effect sizes from a series of meta-analyses in one area, such as classroom-based metacognitive interventions for school-age learners for example, all tend to be between 0.6 and 0.8, and all of those in another area, such as individualised instruction, are all between -0.1 and 0.2, then this is persuasive evidence that schools are more likely to find it beneficial to investigate metacognitive approaches to improve learning, rather than focus on individualised instruction. Some underlying assumptions are that the research approaches are sufficiently similar (in terms of design for example) that the studies compared sufficiently similar samples or populations (of school students) with sufficiently similar kinds of interventions (undertaken in schools) and similar outcome measures (standardised tests and curriculum assessments). So, if you think that a meta-analysis of intervention research into improving reading comprehension has a broadly similar set of studies, on average, to a meta-analysis investigating the development of understanding in science, then you might be tempted to see if any approaches work well in both fields (such as reciprocal teaching) or, indeed, don't work well in both fields (such as individualised instruction). The argument here is that, so long as you are aware of the limits of the inferences drawn, then the approach has value. In medicine, developments such as network meta-analysis (see for example, Mills et al., 2013) aim to develop a more systematic comparative framework by analysing both direct comparisons of interventions within randomised controlled trials and indirect comparisons across trials based on a common comparison (such as a standard treatment or placebo). In education, this methodology has yet to be developed, so more basic inferences provide the best evidence we have, especially where we have no studies providing direct comparisons. As Fraser and colleagues (1987) argued 30 years ago:

Effect sizes permit a rough calibration of comparisons across tests, contexts, subjects, and other characteristics of studies. The estimates, however, are affected by the variances in the groups, the reliabilities of the outcome measures, the match of curriculum with outcome measures, and a host of other factors whose influences in some cases can be estimated specifically or generally. Although effect sizes are subject to distortions, they are the only explicit means of comparing the sizes of effects in primary research that employ various outcome measures on non-uniform groups (pp. 151–152).

Chapter Summary

Research in education has struggled to accumulate findings over time and to build a coherent picture about how we can improve educational outcomes for children and young people across contexts, subjects, ages and educational approaches. Using data from meta-analysis provides a basis to accumulate this information and to make comparisons between approaches. These inferences need to be treated cautiously, particularly where differences are small, as there is usually considerable variation within each meta-analysis. At present meta-synthesis offers us the only way to draw such inferences across studies and across syntheses.

Final Thoughts

I was lucky enough to be taught by Professor Carol Fitz-Gibbon on my research master's course at Newcastle University in the early 1990s. Her papers on meta-analysis, which appeared in the *British Educational Research Journal* in 1984 and 1985, were influential in my thinking about how we accumulate knowledge in education for schools. Hattie's (1992) and Marzano's (1998) studies were the first meta-meta-analyses or 'meta-syntheses' I came across as an education researcher in the 1990s. They influenced my thinking about the value of the approach and the importance of getting a big picture of the relationships between findings in different fields. Sipe and Curlette (1996) set out a vision for the development of the methodology as a key tool for research integration. As I have learned more about the approach and come to understand more about the underlying statistical techniques I have become both more sceptical about the precision of some of the inferences that are possible with our current knowledge, but also even more convinced that we need to develop meta-synthesis as a technique to inform policy, practice and further research. The tongue-in-cheek comment by Cyd Charisse at the beginning of the chapter seemed appropriate as a reminder that apples and oranges can be compared and that these comparisons can be tasty. What we have to remember is to make the correct inference at the level of the comparison. Although an effect size or a moderator variable may have limited meaning at the level of an individual meta-analysis, particularly if the number of studies is small, the stability of effect sizes and associations increases at the meta-meta level. These early syntheses were inspirational and provided the background to the development of what evolved into the Sutton Trust Pupil Premium Toolkit, which was then adopted by the Education Endowment Foundation and developed into the online resource, the *Teaching and Learning Toolkit,* to which I turn in the next chapter.

4   The Teaching and Learning Toolkit

Key questions
How was the Teaching and Learning Toolkit developed?
What topics are included?
How does it use 'meta-synthesis'?

Introduction

This chapter provides an overview of the development of The Sutton Trust – EEF Teaching and Learning Toolkit, ('Toolkit') its origins and rationale. It also describes some of the choices we made in its design and development to provide an explanation of its current structure and form. It has proved popular with teachers and schools in the UK, and internationally. However, like all syntheses, the particular approach has strengths and weaknesses which need to be understood to inform any interpretation of what it means.

The Sutton Trust – EEF Teaching and Learning Toolkit

The approaches used in the early meta-synthesis studies described in the last chapter influenced the thinking behind the methodology and the development of the Toolkit, which I led. The website summarises the messages across different areas of education research to help education practitioners in schools to make decisions about supporting their pupils' attainment. It is available as a web-based resource[1] for practitioners. It uses meta-synthesis as the basis for the quantitative comparisons of impact on educational attainment. The aim of the synthesis is to provide advice and guidance for practitioners who are often interested in the relative benefit of different educational approaches, as well as the detail of how to adopt or implement a specific approach. It includes cost estimates for the different approaches to guide spending decisions.

[1] https://educationendowmentfoundation.org.uk/toolkit/

54

Figure 4.1: Janus, the Roman god of doors and gates

The image I have is for research synthesis is that it is like the Roman god Janus (see Figure 4.1), who was able to see both backwards and forwards. Research synthesis is retrospective in the sense that the research it summarises has already been conducted (and in the case of meta-analysis, this may be many years previously), but it is also prospective in the sense that the purpose of synthesis is to inform and may change what we decide to do in the future.

The initial work drew on a database of educational meta-analyses of intervention findings in education which I compiled as part of an Economic and Social Research Council (ESRC) Researcher Development Initiative (*Training in the Quantitative Synthesis of Intervention Research Findings in Education and the Social Science*) between 2008 and 2011. Towards the end of this project, in summer 2010, I was invited to give a presentation to the Department of Education which was attended by Lee Elliot Major, then the research director at the Sutton Trust. I was perhaps too zealous in my use of normal distributions and effect size estimates in the presentation to communicate my ideas clearly to a policy audience, but Lee invited me to talk further and then to undertake an analysis for the Sutton Trust of the meta-analyses I had collected. The focus was the emerging 'pupil premium' policy to support the achievement of children and young people from disadvantaged background for the coalition government and the evidence from interventions in education which might inform policy spending. Our concern was that the initial policy ideas to spend the allocation on smaller classes and one-to-one tuition were not necessarily the most cost-effective approaches in terms of the existing evidence. The report, 'Toolkit of Strategies to Improve Learning: Summary for Schools Spending the Pupil Premium' was published by the Sutton Trust, and contained 21 summaries of research which could help schools decide how to allocate any additional funding for the recently announced pupil premium initiative in England (Higgins, Kokotsaki and Coe, 2011). The range of topics was deliberately chosen to give an indication of approaches which were less effective as well as those with greater average effects. The analysis included an estimate of impact which was based on the effect size converted into months' progress (see below), together with an estimate of cost of the additional financial outlay for schools and an evaluation of the extent and robustness of the evidence. This was all summarised in the style of a consumer guide to the evidence base (Figure 4.2). The cost estimates, though crude, were a unique feature of the initial Toolkit.

The impact estimates drew on the most rigorous estimates available in a descending order of priority in relation to the evidence. First, meta-analyses of randomised trials and well-controlled experiments where any variation in effect, or heterogeneity, is explored and if possible explained. A quality rating ('stars' for the security of the evidence in the initial version,) also included consistency of effects between meta-analyses to achieve the highest rating. If this kind of evidence was not available, an estimate was made based on other quantitative data, such as correlational studies or estimates from single studies. The reason for this is that the Toolkit set out to provide the best estimate available in a particular area, with an estimate of the robustness of the evidence, rather than only report where the evidence is robust and secure. The Toolkit is therefore distinctive in that it aims to present a summary of evidence in areas where research is sparse or even lacking, at least in terms of causal inference, such as about school uniform or performance pay, but where policy and practice are influenced by people's assumptions about what is or what they think should be effective practice.

A number of factors were serendipitous in supporting the acceptance and uptake of the Toolkit in England. The most important of these was the successful proposal from the Sutton Trust and the Impetus Trust to establish the Education Endowment Foundation (EEF) in early 2011 with a founding grant of £125 million from the Department for Education. This was part of a commitment, tactically championed by the then secretary of state for education, Michael Gove, to help raise standards in challenging schools using rigorous evidence, but also

Toolkit to improve learning: summary overview

Approach	Potential gain[2]	Cost	Applicability	Evidence estimate	Overall cost benefit
Effective feedback	+ 9 months	££	**Pri, Sec** **Maths Eng** Sci	☆☆☆	Very high impact for low cost
Meta-cognition and self-regulation strategies.	+ 8 months	££	**Pri, Sec** **Eng Maths Sci**	☆☆☆☆	High impact for low cost
Peer tutoring/peer-assisted learning	+ 6 months	££	**Pri, Sec** **Maths Eng**	☆☆☆☆	High impact for low cost
Early intervention	+ 6 months	£££££	**Pri,** **Maths Eng**	☆☆☆☆	High impact for very high cost
One-to-one tutoring	+ 5 months	£££££	**Pri, Sec** Maths Eng	☆☆☆☆	Moderate impact for very high cost
Homework	+ 5 months	£	Pri, **Sec** **Maths Eng** Sci	☆☆☆	Moderate impact for very low cost
ICT	+ 4 months	££££	Pri, Sec All subjects	☆☆☆☆	Moderate impact for high cost

[2] Maximum approximate advantage over the course of a school year that an 'average' student might expect if this strategy was adopted – see Appendix 3.

Figure 4.2: The Pupil Premium Toolkit in 2011

drawing on the aspirations for President Barack Obama and Arne Duncan's 'Race to the Top' programme. The EEF adopted the Toolkit as a central piece of evidence synthesis which informed its commissioning of new research in schools in England and its communication strategy with schools. The government also announced in 2012 that schools should report on their websites how they allocated their pupil premium allocation, and this encouraged schools to look for evidence to support their spending. In March 2013, the EEF and Sutton Trust were jointly designated by the government as the 'What Works Centre for Educational Achievement'. The aim of the centres is to summarise and share research with beneficiaries, in this case teachers and school leaders, parents and governors, researchers and policymakers. This is to help them to invest in services that deliver the best outcomes for children and young people as well as ensuring value-for-money for taxpayers. From a personal point of view, this designation is something of a double-edged sword. It sets the Toolkit as a more authoritative voice, at least from a policy perspective, but this can change the way that practitioners engage with the resource. It is therefore even more important that it retains its rigour and independence. My reservations are largely to do with the notion of 'what works' and how people engage with this. As I explain later, in Chapter 8, I think 'what's worked' is both more accurate and more helpful.

With support from the recently formed Education Endowment Foundation to develop the methodology, the summaries were conceptualised as a series of integrated 'umbrella reviews' (Grant and Booth, 2009) which could provide a rigorous but accessible summary with a common methodology across the different areas of educational policy, practice and research. Accessibility is a key feature as engagement with research evidence is a significant challenge (Cordingley, 2008; Hemsley-Brown and Sharp, 2003). The level of presentation needs to be sufficiently familiar to encourage acceptance but sufficiently challenging to promote deeper engagement and support changes in practice. The apparent simplicity at the top level can be deceptive as the messages in any specific area are rarely straightforward, so successive levels of detail aim to support deeper engagement. The Toolkit is therefore presented in summary form at the surface level[2] but with further detail on each area available, right through to the effect sizes and abstracts of the meta-analyses and other studies used in its compilation and synthesis (see Figure 4.3). A technical appendix also sets out the rationale and detail of the effect size calculations and conversions (for more details about the Toolkit approach, the systematic

---

[2] https://educationendowmentfoundation.org.uk/toolkit/toolkit-a-z/

search criteria, evidence estimates and synthesis see Katsipataki and Higgins, 2016). This aims to ensure the synthesis is accurate, with the methods and assumptions made transparent.

As noted in the previous chapter, the inspiration for the Toolkit comes from a number of earlier meta-syntheses. This includes Hattie's (1992/ 2009) work in producing a comparative map of educational research findings, but some of the methodology comes more from others such as Sipe and Curlette (1996) regarding having common inclusion criteria (such as a focus on school-age pupils and relying where possible on intervention research with warrant for causal inference) and a systematic and transparent search strategy for meta-analyses, combined with Marzano's (1998) goal of practical utility with categories which are appropriate for schools. The categories used to organise an evidence map like the Toolkit are particularly challenging. The What Works Clearinghouse tends to focus on named programmes. This is useful for further implementation; you just buy in the programme and use it. However, like Hattie, I wanted to be able to say something at the level of pedagogy to support teachers more broadly. This also risks not being sufficiently detailed to be practical and actionable.

The choice to use months' progress instead of effect sizes and stan-dard deviation units was a difficult one. I was reluctant to let go of the apparent precision offered by two decimal places, but Lee Elliot Major convinced me that we needed a common metric which would be easily understood. The resulting estimate of effect is somewhat crude (an estimate of additional months' progress), but this is probably more realistic in terms of the comparability of the data and the limitations in the precision currently possible from pooling estimates across meta-analyses. Some areas of the Toolkit have relatively consistent findings, such as metacognition and self-regulation (see Chapter 5) or phonics (see, for example, the clustering of the meta-analyses in Figure 1.3 in Chapter 1); others have more widely varying estimates from the differ-ent meta-analyses, such as parental involvement (see Figure 7.1 in Chapter 7) or behavioural interventions.

Our intention was to keep the surface level easy to understand but also to provoke a level of interest or challenge which engages a practitioner in looking deeper. This is the rationale for the way that we chose to communicate the findings. The response from teachers and practitioners, and the overall uptake suggests that it has been proven to be helpful. The balance between accessibility and accuracy is a difficult challenge, particularly when it is also important to engage practitioners in the resource in a way which might benefit practice (for a further discussion about the tensions in research communication for use, see Chapter 8).

Figure 4.3: The Teaching and Learning Toolkit in 2017

A key limitation, as noted earlier, is the single focus on tested attainment and not a wider evaluation of other important achievements and outcomes from schooling.

Behind the Toolkit

The next sections provide more detail about the rationale and the main methods and assumptions used in the comparative synthesis of effect sizes in the Toolkit with reference to other similar approaches where there are key differences. A manual with further technical details is also available on the website. The emphasis is on identifying comparative messages from research, so a number of illustrations and examples will be presented. Comparative meta-analysis has an important role to play in educational research, providing information to inform the field, not about 'what works' but as a summary of 'what has worked'. This is an important distinction. The evidence has accumulated over 30 years or so and derives from a range of contexts, often drawing heavily on research conducted in the United States, and represents what has been effective compared with what was usual practice at the time (the 'counterfactual': see Lemons et al. (2014) for an important discussion of this). This affects the basis for, or the warrant of, the claims of the Toolkit. It argues for professional interpretation and evaluation, rather than a simple 'application' of research findings. The US 'What Works Clearinghouse' makes a wider claim to external validity in both its name and for the programmes it endorses, based on the internal validity of its analysis and synthesis process and of the rigour underlying studies. I am less convinced by the move from internal validity to external generalisability in education, however (see, for example, the classic paper by Bracht and Glass, 1968). The lack of replication in education makes this a particular issue (Ahn et al., 2012). With meta-analysis, the increased precision in the average estimate of effect also comes at a cost of decreased clarity about the nature of the comparison. As you average the study impact, you also 'average' the control or comparison conditions. Precision in an estimate is offset by the increased fuzziness of the counterfactual. The more certain you are that something is effective (on average), the less certain you are about precisely what it is better than. The other reason that makes 'what works' inappropriate is effect sizes are based on the average intervention effect. An effect of 0.2 means about 6% of the intervention group have benefited relative to the controls. For an effect size of 0.8 (often misleadingly called 'large') this only rises to about 28%. You would need an exceptional effect size of 3.0 to get impact for 75% of the intervention group (though nearly all of them would now be likely to be above the average of the comparison group).

Toolkit Themes and Inclusion Criteria

The initial themes for the Pupil Premium Toolkit were based on expectations of how schools seemed likely to spend the pupil premium allocation in England when it was first announced. A number of areas were then specifically included at the request of teachers who have been consulted at different stages in the development of the Toolkit. The initial source of studies was a database of meta-analyses of educational interventions developed for the ESRC Researcher Development Initiative as mentioned above. Additionally, repeated systematic searches have been undertaken for systematic reviews with quantitative data (where effect sizes are reported but not pooled) and meta-analyses (where effect sizes are combined to provide a pooled estimated of effect) of intervention research in education in each of the areas of the Toolkit. These searches have been applied to a number of information gateways. A number of journals which specialise in reviews were hand searched. Journal publishers' websites offering full-text searching were also searched for meta-analyses. Relevant references and sources in existing meta-syntheses, such as those discussed in the previous chapter, were identified and obtained where possible. A record of the search strategy used and studies found are kept for each of the Toolkit themes. Other studies found during the search process are also consulted in each area to provide additional contextual information.

The focus of the Toolkit is on programmes, interventions and approaches which benefit children's and young people's attainment at school, and these criteria form the basis of our search and guide us through the systematic review process. Once the meta-analyses for inclusion have been selected we extract a number of pieces of information. These include the pooled effect size, standard error, type of effect size (Hedges' g, Cohen's d or Glass' Δ: see Appendix B), confidence intervals, number of studies in the meta-analysis, whether there is a moderator analysis and publication bias, among others. In each area of the Toolkit an overall estimate of the effects is then identified. Where the data are available and suitable a weighted mean is calculated. This is based on calculating a weight for each meta-analysis according to its variance (based on the reciprocal of the square of the standard error: Borenstein et al., 2009). Where the data are not available for this an estimate is given based on the available evidence (such as mean and median effects) and a judgement made about the most applicable estimate to use (such as the impact on disadvantaged pupils, or the most rigorous of the available meta-analyses). Where no meta-analyses of educational interventions in a given area can be found, then an effect size is estimated from correlational studies or large-scale studies investigating the relationship under review. If there is no information from any of these types of research can be found, then individual studies are identified which can provide a broad estimate of effect. The priority during the systematic review is to find rigorous meta-analyses, but in areas where this is not possible (mostly due to lack of available research) there are identified quality criteria that are assigned to each strand to depict the quality of evidence for each topic. Padlock ratings are provided for each strand as a 'security' classification. This acts as a useful guideline providing information about how extensive and robust the evidence is. Comparisons can be made about the strength of the evidence in each area, the number and quality of meta-analyses and the reliability of the impact across each strand of the Toolkit. Hence, a more complete picture can be drawn regarding the impact of each approach as this is important to bear in mind when making decisions. For example, aspiration-based interventions currently have little to no impact on attainment, but this is only one part of the picture since the evidence base for this area is weak with no meta-analyses available, as noted in two recent high-quality reviews conducted for the Joseph Rowntree Foundation (Gorard et al., 2012; Cummings et al., 2012). Before jumping to conclusions, we need to consider all of the available information. No evidence of effect, based on extensive evidence is very different from no evidence. Again, this is an important feature that can be incorporated in relevant summaries. By contrast, the US 'What Works Clearinghouse' only reports outcomes from studies which meet its rigorous inclusion criteria. This is important when summarising effects of programmes where there is extensive evidence and when the question is whether a particular intervention works or not. But by excluding areas with little rigorous evidence there is a danger that areas of interest to teachers and schools may be overlooked. Whilst the level of confidence in less rigorous research is lower, it is often a useful indicator and can provide more indicative evidence. Another consideration is that correlational and descriptive research is often essential to build new ideas and theories that can be further investigated using experiments or explore areas where experimental designs are not possible (Slavin, 2002). We do recognise the value of well-designed research, but we also suggest that areas that have educational importance should not be overlooked only because there are no or limited rigorous evidence from randomised controlled trials. On the contrary, a more complete perspective should be presented for the practitioners to use this information to get a better understanding of different research areas.

Effect Sizes: Presentation, Interpretation and Implications

As discussed in Chapter 1, an effect size is a key measure in intervention research and an important concept in the methodology of the Toolkit as well as a common metric used to present meta-analytic findings. It is basically a way of measuring the *extent* of the difference between two groups. It is fairly easy to calculate and can be applied to any measured outcome for groups in education or in other areas of research more broadly. For the Toolkit, the kind of effect size used is the standardised mean difference. For a discussion of other types of 'effect size' see Appendix B.

The value of using an effect size is that it quantifies the effectiveness of a particular intervention, relative to a comparison group. It allows us to move beyond the simplistic, 'Did it work (or not)?' to the far more important, 'How *well* did it work across a *range* of contexts?' It therefore supports a more critical and rigorous approach to the accumulation of knowledge by placing the emphasis on the most important aspect of the intervention – the size of the effect – rather than its statistical significance, which conflates the effect size and sample size and requires a number of assumptions, such as random sampling, which are not usually met in education. For these reasons, effect size is the most important tool in reporting and interpreting effectiveness, particularly when drawing comparisons about *relative* effectiveness of different approaches. A perhaps under-used benefit of using effect size is that findings can also be converted back to the original scale, whether that is reading progress or national examination scores. We have not attempted to do this in the Toolkit, as the focus is on emphasising the relative benefits of different approaches. However, it would be possible to do so, and this would provide a more specific estimate of impact on a test or outcome of interest to a practitioner.

In the Toolkit, we were keen to promote the understanding of the effect size to different audiences and to make the findings more accessible. Therefore, we equated effect size to a single scale of school progress in months as a crude but meaningful equivalent. The use of a single scale which was intuitively easy to understand was an important aspect of the accessibility of the entry level of the overall Toolkit summary. We have estimated that a year of progress is about equivalent to one standard deviation per year and corresponds with Gene Glass's observation that 'the standard deviation of most achievement tests in elementary school is 1.0 grade equivalent units; hence the effect size of one year's instruction at the elementary school level is about +1' (Glass et al., 1981: p. 103). However, it is important to note that the correspondence of one standard deviation to one year's progress can vary considerably for different ages and types of test. It is also the case that effect size difference reduces with age. Hill and colleagues (2008) estimate annual progress on tests drops from 1.52 to 0.06 for reading and from 1.14 to 0.01 for mathematics in the US from kindergarten to grade 12. Wiliam (2010) estimates 'apart from the earliest and latest grades, the typical annual increase in achievement is between 0.3 and 0.4 standard deviations'. One implication of this is that our estimates of improvement may underestimate the gains achievable for older pupils. By the end of secondary school age, the difference between the attainments of successive age groups is relatively small, especially compared with the spread within each. For these older pupils, it may be misleading to convert an effect size into typical months' gain. One month's gain is typically such a small amount that even quite a modest effect appears to equate to what would be gained in a long period of teaching.

There are other reasons for preferring a more conservative estimate of what it likely to be achievable in practice. One problem is that estimates of the effects of interventions come from research studies that may optimise rather than typify their effects. Another is that average effects for a class may represent progress for a relatively small number of children in that class. For these reasons, it may be unrealistic to expect schools to achieve the gains reported in research whose impact may be inflated (this is what Cronbach and colleagues (1980) called 'super-realization bias'). Other evidence suggests that effect sizes will also be smaller as interventions are scaled up or rolled out (Wigelsworth et al., 2016). Slavin and Smith (2009) report that there is a relationship between sample size and effect size in education research, with smaller studies tending to have larger effect sizes (see Chapter 8 for a further discussion of this). This may be due to the stage of intervention (pilot, efficacy and effectiveness) as Wigelsworth and his colleagues have shown.

A further problem is that part of the learning gain typically achieved in a year of schooling may be a result of maturational gains that are entirely independent of any learning experiences that are, or could be, provided by the school (Luyten et al., 2006). Researchers and practitioners should therefore take into consideration the potential variables that can affect the observed gains in pupils' progress. For these reasons, we have also selected what we see as a more conservative estimate, based on effect size estimates for younger learners, which can be improved or refined as more data becomes available about effect size transfer from research studies to practice.

Presenting effect sizes as month's additional gain has been proven to have had a significant role in the communication of our findings to practitioners, schools and people who do not have (and may not want) a full understanding of effect sizes. Presented in a simple but meaningful way, research findings are disseminated to diverse audiences. We believe that this is an important aspect of our approach, especially for educational research, though there is a trade-off here again between accuracy and accessibility. One of the main criticisms that we encountered during the development of the Toolkit was that frequently educational research lacks clarity in presenting research findings to wider non-academic audiences. Thus, translating the effect size into month's gain made results from meta-analyses and quantitative studies easy to comprehend but at the cost of some of the precision in these estimates and its applicability at different ages and stages of education.

There are some further notes of caution in comparing effect sizes across different kinds of interventions and evaluations (see Cheung and Slavin, 2015). Effect size as a measure assumes a normal distribution of scores. If this is not the case, then an effect size might provide a misleading comparison. If the standard deviation of a sample is decreased (e.g., if the sample does not contain the full range of a population) or inflated (e.g., if an unreliable test is used), the effect size is affected. A smaller standard deviation will increase the effect size whereas a larger will reduce it. Another key issue is which standard deviation is chosen (Hill et al., 2008) as this primarily determines the comparability of the effect size (Coe, 2002). This choice can explain the variation in methods advocated above. For example, a decision has to be made as to whether to use the control group's standard deviation or a 'pooled estimate' of both the experimental and control group. There are different implications involved in the above choice, described in detail by Olejnik and Algina (2000).

There is also some variation associated with the type of outcome measure with larger effect sizes typically reported in mathematics and science compared with English (e.g., Higgins et al., 2005) and for researcher designed tests and teacher assessments compared with standardised tests and examinations (e.g., Hill et al., 2008: p. 7). Finally, studies without randomisation which focus on learners from either end of the distribution (high attaining or low attaining learners) are likely to be affected by regression to the mean if they do not compare similar groups or learning levels (Shagen and Hogden, 2009). This could inflate effect sizes for low attaining pupils (who are more likely to get higher marks on re-test) and depress effect sizes for high performing students when they are compared with 'average' pupils. If the correlation between pre-test and post-test is 0.8, regression to the mean may account for as much as 20% of the variation in the difference between test and re-test scores when comparing low and average students.

In summary, considerable caution is needed in making comparisons where a number of factors need to be taken into account in understanding what influences effect size estimations. One of the assumptions in the Toolkit, as in Hattie's *Visible Learning*, is that the overall distribution of studies in educational research is sufficiently similar that the patterns which emerge from this analysis represent our best indicator of the relative effects of different approaches. An important part of the defence of this approach is that I think this is an empirical question which we can explore. In the Toolkit, we have focused on intervention studies for two reasons. The main one is that we are advocating that schools introduce new approaches, which is more similar to an intervention. Any benefit is likely to be similar. The second is that I consider that effect sizes are more likely to be comparable across interventions (as opposed to comparisons across a wider range of research designs). I think the current approach to meta-synthesis is the best that we have, but it could be refined in two ways, first by undertaking meta-analyses which are designed to be more compatible (as Sipe and Curlette argued in 1996), such as by having common inclusion criteria and comparing impact on similar groups of pupils with similar kinds of outcome measures and undertaking similar moderator or regression analyses. The second is by looking at how good a predictor the findings from meta-analysis are for the findings of future studies to see whether they follow a similar pattern. Between 2011 and 2017, the EEF commissioned more than 145 experimental trials involving nearly a million pupils in England, with one aim being to feed the results back into the Toolkit to test how useful the findings from previous research are as predictions of impact in subsequent interventions. A third way would be to combine at the study level, rather than pooling meta-analyses. This would need a single database with all of the studies in the Toolkit (there are currently over 10,000) to create a single mega-meta-analysis. The comparability of effect sizes could then be explored in more detail, and more precise estimates for different subjects and ages of learners could be made (for a further discussion of future possibilities see Chapter 8).

Cost-effectiveness Estimates

A key element of the Toolkit is the importance of the cost–benefit analysis that is included for each intervention. The argument here is that an approach which is cheap and easy to adopt, but has a relatively low impact, may still be better than one which is very expensive, even if it has a larger impact. This is implicitly a criticism of Hattie's *Visible Learning,* where he argues that schools should look for approaches that work better than the average effect size of 0.4, which he identifies as a 'hinge' point. The Toolkit focuses, however, on intervention research where the progress is additional to the control or comparison group's progress, which includes maturational effects. Cost estimates in the Toolkit are calculated based on the additional likely costs of adopting an approach with a class of 25 students. They do not take into account existing teachers' salary costs or the overall running costs of the school. This was a pragmatic decision which relates to the Toolkit's origins in the pupil premium policy in England where schools had to justify their expenditure in tackling disadvantage. The cost estimates are therefore for the additional outlay needed to implement an approach or intervention. In cases where an approach does not require any additional resources, estimates are based on the cost of training or professional development to establish new practices. In terms of cost-effectiveness there are of course other issues to consider. For example, reducing class sizes only lasts for as long as the funding maintains smaller classes. Technology equipment typically lasts for three to five years. On the other hand, developing teachers' skills through professional development can be more valuable since it may develop capacity and enable longer lasting changes within the school environment.

This feature of the Toolkit has also been proven to be valued, informing school leaders and teachers on this important dimension upon which, among others, they base their decisions on how to spend the Pupil Premium (in England) or other discretionary spending. Educational research rarely reports the costs of different approaches, but from both a policy and practice perspective this of crucial importance. This makes the Toolkit distinctive when compared with other approaches to evidence synthesis in education. As with the translation of effect sizes into months' gain, the cost estimate is a simplification, again reflecting the tension between accessibility and accuracy when summarising findings from research.

Examples of Comparative Meta-analysis from the Toolkit

In 2017, at the time of writing, the Toolkit comprised 34 topics summarising different approaches to intervention having as a main goal the improvement of learners' attainment in school. Each topic has information about how much the intervention costs to implement, the security of the evidence and how many month's gain can be achieved. For each strand, more detailed information is provided about the approach, with a definition, further information about the impact, the security of the evidence, the costs, and questions to consider for practitioners (based on what is associated with more and less successful impact from the moderator analyses in the meta-analyses). The top level of online presentation of the Toolkit provides both a brief and quick overview of the topics as well as a more thorough presentation at the next level for those who want additional detail. The tension here is in presenting comprehensive information and being succinct. All of the information on which the analysis has been made is available in a technical appendix for each strand at the third level.

The idea is that schools considering different approaches to support learning can consider the strengths and limitations of the different approaches. Mentoring might have a low impact on attainment but it is cheap to administer, so compared with one-to-one tutoring, which is more expensive, teachers might decide to try mentoring. Another example having different comparisons is reducing class sizes and one-to-one tutoring. Both have high costs but, on average, the former has three months' gain and the latter five. One-to-one tutoring looks like a better bet, especially as it can be targeted. Another variable that can be considered is the evidence rating. Small group tuition has a limited evidence rating, providing four months' additional gain. On the other hand, summer schools have two months' additional gain in attainment but more extensive evidence of effectiveness. Therefore, teachers might consider using an approach which appears more promising based on the extensiveness of the evidence surrounding this approach and may have some ideas about how to mitigate the challenges (such as attendance at summer school by those who most need it). As argued above, synthesis of evidence in terms on impact, cost and evidence strength should all inform schools' decision-making, not as guarantees of 'what works' but as indicating the likely chances of success. If a school chooses an area where effects tend to be lower and the evidence is robust, then we suggest that they need a clear understanding of what they should do to ensure they are more effective than the average approach described in the studies summarised in meta-analyses. If they choose an area of high impact, they will only need to be as successful as researchers and schools typically were in the studies included in the synthesis: a 'good bet', as opposed to a riskier one.

What Is It That 'Works'?

This perspective also encourages schools to take responsibility to ensure that new approaches are successful, rather than assuming that it is something which 'works' without this purposeful commitment. I've always struggled with the idea that an intervention or an approach 'works'. At best, an intervention or approach is a structure or a catalyst which reliably helps people do things differently, to act and behave in different ways than they would have otherwise. An intervention or approach is a tool, built of ideas and resources. It does not have agency and intent. I accept it is very cumbersome to say more explicitly what we mean. 'Research shows that teachers can successfully train pupils to use a reciprocal teaching strategy to interrogate texts as they practice their reading which, on average, improves these students' comprehension' is a much longer, if more accurate, way of saying 'reciprocal teaching improves reading comprehension', or even 'reciprocal teaching works'. The shorthand version is quicker and easier, but this airbrushes the teacher, the learners and their interactions from the picture and obscures the actions that they need to take for the use of reciprocal teaching to be successful. Who is it that is doing the 'work', the intervention, or the teacher and the learners?

A Further Aside about Statistical Significance

The idea behind 'what works' is that research has proved that it is effective. However, the assumptions needed for null hypothesis testing are problematic. Remember the convoluted logic and arbitrary levels of likelihood invoked by null hypothesis testing outlined in Chapter 1? There are two further assumptions needed to make the mathematics work. You need a random sample of the population of pupils before you start your experiment, and this population should be normally distributed. Here there is a glimmer of hope as most large samples of children are normally distributed on tests if the tests are good enough. Anyway, if you conduct your experiment, reject the null and accept the alternative hypothesis you are in effect saying something like this: 'I think the difference I found between these groups would be unlikely to occur more than one in 20 times by chance, had we not expected to find any difference and had I randomly selected my sample from the population (but I was expecting to find a difference and I didn't randomly select my sample from a defined population). So, ignoring these assumptions or at least believing the analysis to be robust to these assumptions not being met, I conclude that there really is a difference between the groups and that therefore this intervention was successful.' One final issue is also important to consider: How likely is it that we could have made a mistake? This has nothing to do with the p-value we just worked out. The p-value is the probability of the data occurring, assuming the null hypothesis with random sampling and a normal distribution. The p-value is not the probability of the alternative hypothesis being true. In fact, using this logic we are likely to make a mistake at least 23% of the time and possibly as much as 50% of the time (Sellke et al., 2001) and incorrectly reject the null (I hate these double negatives!), 'Ah', I ask finally, 'so the difference was statistically significant, given the convoluted logic and the assumptions which I hope are sufficiently robust to non-random sampling, and there is still a worrying chance we've got it wrong, but is it actually educationally important?' Here we have to turn back to the effect size and make a judgement based on the extent of the difference in relation to the confidence intervals. There are other approaches based on other mathematical assumptions, such as Bayesian methods, which in most circumstances produce very similar results (Xiao et al., 2016), suggesting the assumptions of the frequentist approach are robust in most of the educational trials we have analysed. Personally, I still find a p-value informative (particularly when it is tiny), but I think of the confidence intervals that it generates as the minimum sampling uncertainly, had you selected a random sample of the population (when you haven't). It is likely that there will be other issues with the research, such as weaknesses in the design or problems with test reliability or attrition, which should worry us far more in deciding if something 'worked'.

A Living and Growing Review

The Toolkit is unusual in that the web version allows regular updating (we aim to review each strand each year). This lets us pick up new meta-analyses and single studies to re-estimate the overall effect and adjust the text to reflect new research and understanding. The idea of a 'living review' is not new, and part of the aim of a systematic review is that it can be updated, without having to start from scratch (Elliot et al., 2014). What makes the Toolkit different is that the configuration of strands can also be changed. We can add or delete themes to reflect new evidence or to meet a need in changing practice.

Chapter Summary

The Sutton Trust – EEF Toolkit is an example of meta-synthesis which aims to build on earlier work in this field. The overall aim of the Toolkit is not to provide definitive claims as to what will work to improve learning as a guarantee of future success. Rather it is an attempt to provide the best possible typical estimate of what is likely to be beneficial based on existing evidence. It exemplifies how I think this kind of analysis should be used to inform practice. More specifically, it serves as an example of the different aspects of research which are needed in communicating findings in the educational arena, whilst also making evidence more accessible for practitioners and decision-makers.

The variation in findings in education and the aggregation process means that applicability of this information to a new context is always going to be a probability rather than a certainty. Meta-analysis emphasises the average effects, rather than the range of impact and what might cause any variation. Application is always likely to need interpretation and active enquiry or evaluation to ensure it achieves the desired effects. This requires professional judgement and commitment to engaging with evidence but also a disposition to interpret, challenge and test particular findings to ensure they are helpful.

Final Thoughts

Summarising evidence is a set of tensions between accuracy and accessibility and between the general and the specific: it looks back to look forward. The image at the beginning of the chapter is of the Roman god, Janus. As mentioned in the chapter introduction, this is because for me he symbolises the dual retrospective and prospective nature of research synthesis. We undertake synthesis to inform what we do in the future (whether as education researchers or practitioners). This is a key part of the challenge, we look back with a wide glance across the available research and use this to direct where we go next. The view behind is the same for everyone, but the next step will depend on exactly here you want to go. There will not be one best approach as there are many potential users of any summary or synthesis, each with their own specific needs or goals. Any approach to synthesis will need to balance these differing needs. I've learned many lessons through the development of the Toolkit, from colleagues in academia and at the Sutton Trust and the Education Endowment Foundation and from the teachers and school leaders I've presented, discussed and tried to apply the findings with. We've tried to build on what the pioneers in the field achieved and tried to focus on what is helpful for teachers and schools. There is still much to be done, but the view of the research landscape is reasonably consistent and coherent, which should now help in being clearer about the next steps needed to develop the use of evidence from research in education.

Part II

What's Made a Difference to Learning?

The next section forms the second half of the book and considers the findings from meta-analysis in different areas of education research. This is to exemplify some of the challenges in using this approach to quantitative synthesis to draw conclusions about effective educational practice. Chapter 5 summarises some of the findings from meta-analyses about pedagogy and about effective approaches to teaching and learning. The evidence about feedback, metacognition and self-regulation and the impact of digital technologies on learning are reviewed in turn. The chapter concludes with a review of the perplexing evidence about learning styles approaches in education. Chapter 6 reviews the evidence on four areas of literacy intervention: early years, speaking and listening (oracy), reading and then writing, so as to return to some of the themes used to introduce the book. Chapter 7 draws on the evidence about parental engagement and involvement with schools. This is to consider the challenge of drawing conclusions for practice in schools across meta-analyses. Developing effective partnerships to improve outcomes for children is not straightforward. Chapter 8 draws together the argument of the book as a whole about the value of meta-analysis and meta-synthesis as a technique to understand the evidence about effective teaching and learning, its limitations and potential.

Each chapter is again headed by a number of key questions which are addressed in the course of the chapter. Some sections have a more academic presentation, but I have included my personal reflections to draw out what I see as the issues of interpretation in relation to the use of meta-analysis.

Chapter 5: Meta-analysis and Pedagogy
    What patterns of effects appear when comparing
    meta-analyses?
    Are there general implications for teaching and learning?
    What can we infer from specific areas such as feedback
    and metacognition?

Chapter 6 was informed by the work developing the literacy guidance for the EEF's North-East literacy campaign (EEF, 2017). Chapter 7 has been developed from an article I wrote with Maria Katsipataki in 2015 for the *Journal of Children's Services* entitled 'Evidence from Meta-analysis about Parental Involvement in Education which Supports Their Children's Learning' (Volume 10.3: 280–290). I am grateful to Emerald Publishing Limited for permission to reuse and develop this material. Chapter 8 has a number of influences and I should acknowledge two research projects in particular. These are *Closing the Gap: Test and Learn for the National College* (Churches, 2016; Childs and Menter, 2017) and *Evidence-Informed Teaching: An Evaluation of Progress in England for the Department for Education* (Coldwell et al., 2017). I have also benefited from personal discussions at various times with Professor Gert Biesta and Professor Nancy Cartwright. Their ideas, writings and questions have honed my own thoughts on issues relating to evidence and education developed in Chapter 8.

5   Meta-analysis and Pedagogy

Key questions
What patterns of effects are evident when comparing
  meta-analyses?
Are there general implications for teaching and learning?
What can we infer from specific areas such as feedback
  and metacognition?
What do meta-analyses of digital technology and learning
  styles tell us?
What challenges arise in using evidence in this way?

When I was a kid about half past three
My ma said 'Daughter, come here to me'
Said things may come, and things may go
But this is one thing you ought to know ...
Oh 't ain't what you do it's the way that you do it
'T ain't what you do it's the way that you do it
'T ain't what you do it's the way that you do it
That's what gets results

T'ain't What You Do (It's the Way That Cha Do It) Oliver
and Young, 1939

Introduction

This chapter summarises some of the findings from meta-analyses
about pedagogy and about effective approaches to teaching and
learning. I will look at areas such as feedback and approaches
which encourage learners to thinking about and take responsibility
for aspects of their learning (metacognition and self-regulation).
I want to provide an overview of some of the patterns in the findings
to set these kind of research findings in a broader context.
The 'Digital Technologies' section looks at the impact of digital
technologies on learning to consider what it is that makes
a difference. I conclude with a review of the evidence about learning
styles, which is a problematic area of research for meta-synthesis for
a number of reasons. Each of these different areas is used to

77

exemplify some of the key challenges in drawing conclusions from this kind of evidence in education in terms of conceptual coherence, what we mean by impact and the importance of trying to understand the underlying causal mechanisms responsible for any improvement in teaching and learning.

The Importance of Pedagogy

Looking at the findings across the Toolkit and other similar syntheses provides opportunities to look at the broader implications in terms of the patterns of effects across meta-analyses. One of these is the relative benefit of approaches which focus on the quality of interactions in teaching and learning processes, as opposed to those which focus on structural features in education. The closer you are to the teaching and learning process, or the greater the focus on the interactions between teachers and learners and between learners themselves, such as feedback and peer collaboration, the more likely you are to see larger effects in learning outcomes.

Proximal and Distal Influences

Interventions which directly influence these teaching and learning interactions, such as feedback, metacognition and self-regulation, peer tutoring, one-to-one tuition and collaborative learning, are grouped as the upper end of effects (between 0.4 and 0.7). At one level this is not surprising, and these are often described as proximal effects due to the proximity of the intervention to the outcomes. By contrast, more distal influences on teaching and learning such as performance pay, class size, ability grouping or school uniforms have much lower average impact (effects between 0.01 and –0.09). Again, this is not surprising as there is not necessarily a direct influence from these approaches on teaching and learning. The very small average effects also argue for caution, in that when people aim to improve outcomes by focusing on more distal approaches in education they are rarely successful. This indicates that if schools are interested in improving outcomes directly, then focusing on teaching and learning is likely to be much more productive (see Figure 5.1). This is similar to Seidel and Shavelson's (2007) analysis of teacher effectiveness research where similar proximal and distal effects have been noted.

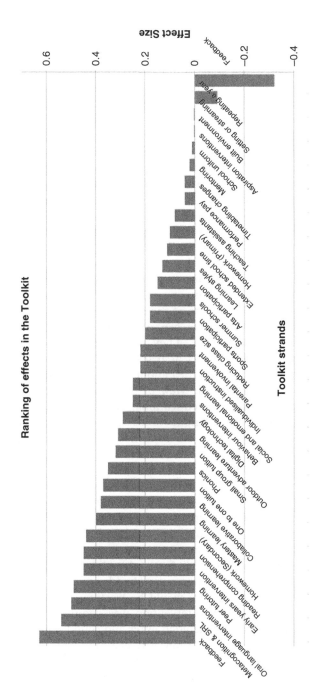

Figure 5.1: Toolkit approaches ranked by effect size

Treatment Inherent and Treatment Independent Measures

However, it is also possible that some of the greater impact is related to how closely the intervention was related to the outcome measures used to assess the effects. It is not surprising that a metacognition intervention improves metacognition; what we need to know is whether it helps with learning at school. A metacognition measure would be described as 'treatment inherent' in a metacognition intervention whereas a standardised test of reading comprehension or mathematics for a similar intervention might be considered 'treatment independent'. This issue is a key challenge for meta-analysts. One of the benefits of meta-analysis is that it lets us compare the impact across different outcome measures. It standardises the gain by the spread (the difference in means divided by the standard deviation: see Chapter 1). However, this means that you can combine different measures which may relate to different aspects of the intervention or even combine outcomes which are inappropriate. It is like making a smoothie from research findings. One of the challenges with experimental research design is to avoid measures which are treatment inherent (Slavin and Madden, 2011). They may give you an idea of how successful the programme was at teaching specific skills or content and may also give you a good idea about an aspect of fidelity, but this might give an over-optimistic view of the actual impact on learning. Most fidelity measures in educational research are input measures (such as number and length of sessions delivered or a record of attendance). A treatment inherent measure can give you an output measure of fidelity, as it can confirm that the intervention successfully altered what learners did.

Consider a phonics intervention in the early years which teaches letter sounds and letter recognition to young children. If the intervention group does better than the control group on letter and sound recognition at the end of the intervention, this would not be particularly surprising, especially if the control or comparison group had no formal literacy teaching during the intervention period. In fact, if they did not do better you would reasonably wonder if the intervention had been successfully delivered. In this instance, the measure would be treatment inherent. In another trial, where the phonics intervention was compared with the existing literacy teaching, which included letter and sound recognition, it might be a fairer assessment. However, even in this second example you still need to be careful about how you interpret the outcomes. The intervention group may perform better on letters and sounds at the end of the intervention, but the other group may have been successful at teaching or developing other valuable literacy skills and knowledge which were not tested, and as a result, the group that received the existing literacy teaching will catch up, or even be more successful, over time. Ideally, for both you want to wait to see whether or not children become more successful readers in the long run. Otherwise the intervention might just be like squeezing a balloon and measuring the diameter at the bulge. When you let go the balloon springs back into shape and loses this girth. Educational interventions can be like this; you develop particular skills and knowledge so that at post-test it appears that the intervention group has improved. However, very soon the impact disappears, and there is no identifiable long-term benefit.

It is certainly true that the effects of most educational interventions tend to fade over time, and we have relatively few studies which look at the long-term outcomes (Protzko, 2015). The typical fade out is initially fairly rapid but shows a geometrical decline (Li et al., 2017/in preparation[1]), with about half of the advantage lost in the first year then a flattening out to an average of about 0.01 SD per year. What is hard to determine is when there were real effects which have faded and which were illusory or inflated because they were treatment inherent. (For a further discussion of the loss of impact over time see Chapter 6.)

Feedback: Easy to Say, Hard to Do Well

Feedback is ranked at the top of the Toolkit strands in terms of impact. It sounds deceptively simple. You just tell a learner where they have gone wrong, and they will improve, won't they? However, this simple idea is deceptively complex. Feedback is often defined as information given to the learner and/or the teacher about the learner's performance relative to learning goals. It should aim towards (and be capable of producing) improvement in children's and young people's learning. Feedback redirects or refocuses either the teacher's or the learner's actions to achieve a goal, by aligning effort and activity with an outcome. You teach something, you see what happens and you use this information to refocus what you do next, according to your objectives. What has

---

[1] I am very grateful to Professor Greg Duncan for sharing a working draft of the paper by Li, W., Duncan, G. J., Magnuson, K., Schindler, H. S., Yoshikawa, H. and Leak, J. (2017/in preparation) 'Timing in Early Childhood Education: How Cognitive and Achievement Program Impacts Vary by Starting Age, Program Duration, and Time Since the End of the Program'.

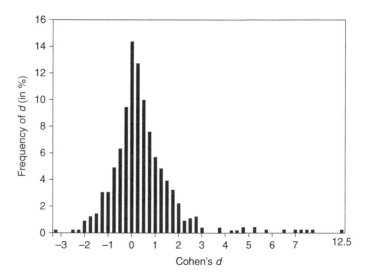

Figure 5.2: Distribution of 607 effects of feedback intervention on performance (from Kluger and DeNisi, 1996)

happened provides input to create a feedback loop and makes teaching contingent upon learning. So far so good. However, this is where it gets challenging. Feedback can be about the learning activity or task itself, about the process of undertaking the activity, about learners' management of their actions or their learning or their self-regulation or even (the least effective) about learners as individuals ('good girl' or 'clever boy': see Hattie and Timperley, 2007). This feedback can be non-verbal or verbal as well as ephemeral (an encouraging nod) or more permanent (e.g., written). It can come from a teacher or someone taking a teaching role or from peers or automated through technology or even from the learner themselves. It can be direct or indirect and implicit. It can be immediate or delayed. Where aspects of feedback have been researched, the studies tend to show very high average effects on learning. However, there is also has a very wide range of effects and some studies show that feedback can have negative effects and make performance worse (see the distribution of effects from Kluger and DeNisi (1996) in Figure 5.2). Bearing in mind the important observations about average treatment effects (see Chapter 8) each of these studies represents an extremely wide range of effects on the participants involved. Figure 5.2 under-represents this variation at an individual level. It is therefore important to understand the potential benefits and the possible limitations of using feedback to help learning.

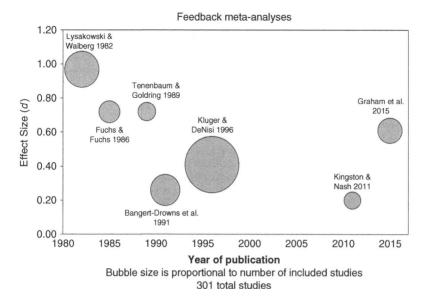

Figure 5.3: Bubble plot of feedback meta-analyses by year

In general, research-based approaches that explicitly aim to provide feedback to learners, such as Bloom's 'mastery learning', also tend to have a positive impact. You could think of peer tutoring and peer assessment as feedback interventions, and certainly feedback is an important component. Feedback has been documented to have effects on all types of learning across all age groups. Research in schools has focused particularly on literacy, mathematics and, to a lesser extent, science. One concern I have about the evidence for feedback is that the older meta-analyses tend to show higher effects (see Figure 5.3; further details about these meta-analyses can be found in Appendix A.5.1).

Research evidence about feedback was part of the rationale for Assessment for Learning (AfL) in the UK. One evaluation of AfL indicated an impact of half of a GCSE grade per student per subject is achievable, which would be in line with the wider evidence about feedback. Other studies reporting lower impact indicate that it is challenging to make feedback work effectively in the classroom. This has also been demonstrated in a recent EEF pilot study where teachers tried to apply the evidence on feedback through an action research approach.

Table 5.1: *Feedback meta-analyses*

Meta-analysis	Focus	Number of studies (k)	Pooled effect
Lysakowski and Walberg (1982)	Instructional effects of cues, participation and corrective feedback	54	0.97
Fuchs and Fuchs (1986)	Formative evaluation on achievement	21	0.72
Tenenbaum and Goldring (1989)	Cues, participation, reinforcement and feedback and correctives on motor skill learning	15	0.72
Graham et al. (2015)	Formative assessment and writing	27	0.61
Kluger and DeNisi (1996)	Feedback interventions on performance	131	0.41
Bangert-Drowns et al. (1991)	Feedback in test-like events	40	0.26
Kingston and Nash (2011)	Assessment for learning	13	0.20
Total studies and overall weighted mean effect size		*301*	*0.63*

There are a substantial number of reviews and meta-analyses of the effects of feedback (see Table 5.1). The idea is conceptually clear and coherent, if complex. Educational (rather than psychological or theoretical) studies tend to identify positive benefits where the aim is to improve learning outcomes in reading or mathematics or in the recall of information or on performance in learning tasks. Later meta-analyses have tended to focus on the application of feedback in teaching and learning contexts. A meta-analysis of studies focusing on formative assessment in schools (Kingston and Nash, 2011) indicates the gains are typically more modest, suggesting an improvement of about three months' additional progress is achievable in schools or nearer four months' when the approach is supported with professional development. Graham and colleagues' meta-analysis (2015) focuses on formative writing assessments that were directly tied to everyday classroom teaching and learning and suggests feedback in classroom contexts can improve pupils' writing performance.

Feedback: A Goldilocks Problem

A 'perfect' model of feedback for a teacher is difficult to define. This is because it is a relational set of decisions dependent upon the teacher, the learner and what is being taught in a specific context. At a simple level, good feedback leads to better learning, ideally immediately, but certainly over time. It is a Goldilocks problem, in which the challenge is to get it 'just right'. The teacher responsible for providing the feedback (which includes managing learners so they get feedback from each other or from another source) has a current level of knowledge and expertise, such as in terms of their general teaching skills, their knowledge of the content being taught (both subject and pedagogic knowledge) and their current knowledge of the learner (such as their progress and their previous receptivity to feedback). This makes judging the precise nudge to give through feedback complex and nuanced, and also very specific. Generic strategies risk missing the precision needed, but unless you understand the range of possibilities from different strategies it is likely that you miss the opportunity for the optimal nudge.

Feedback in schools is often public so must also be understood in relation to how it is received as well as how it is given or intended. A trusted or respected teacher will have more impact than one where the pupil is more interested in the reaction of her peers. I think of it as like pushing a swing. It looks deceptively simple but is a coordinated action requiring accurate timing, force and direction: too gentle and the learner slows down, too hard and they fall off!

Thinking for Yourself and Managing Your Own Thinking

Metacognition and self-regulation approaches (sometimes known as 'learning to learn') aim to help learners think about their own learning more explicitly and to take increasing responsibility for their own learning as part of the learning process. This is usually by teaching specific strategies to pupils to set goals and monitor and evaluate their own academic development, in a specific curriculum context, through specially designed tasks. Self-regulation is often a broader term and also means managing one's own feelings and motivation regarding learning. The intention is often to give pupils a repertoire of strategies to choose from during learning activities and to encourage them to take responsibility for using these strategies effectively. Learning to learn can easily be misunderstood as teaching general strategies and study skills without reference to specific content. The researched interventions tend to have a very specific subject context with a clear focus, and they involve the learning of specific skills and knowledge. Metacognition and self-regulation aim to improve the efficiency of learning either immediately in the short term or as a more strategic educational goal so that over time the learner takes increasing responsibility for their own learning.

Much of the early work in metacognition in the 1970s was done in two fields, memory (recall) and problem-solving. Flavell, who coined the term, saw similarities in the difficulties learners had in managing their own thinking between these two areas and identified the value of focusing on metacognitive knowledge and metacognitive experiences. Even now the terminology can be confusing, with psychologists often focusing on self-regulation, educationalists on metacognition and then neuroscientists also talking about executive function. All refer to aspects of the way the individual manages their own thinking and performance, but with a difference in emphasis and focus.

Metacognition and self-regulation approaches have consistently high levels of impact, with classes making an average of eight months' additional progress across the meta-analyses in the Toolkit (see Table 5.3); or for every 100 pupils taught about 20 will benefit compared with business as usual. The evidence indicates that teaching these strategies can be particularly effective for low-achieving learners and for older pupils. However, you have to be metacognitive about a specific area of cognition. This can be represented as a being 'meta' about different kinds of thinking. You can be metacognitive about relatively simple aspects of thinking such as remembering and recall. You can think of different strategies to help you remember, either to retrieve information you know that you know, or you can think of and use strategies to help you remember in the future (e.g., mnemonics or Sherlock's 'mind palace'[2]). This is like deliberately controlling or managing your own thinking. You can also be metacognitive about more complex thinking or more complex tasks such as problem-solving using mnemonics like RIDE (Read Identify Determine Enter (the numbers)) or using creative thinking strategies by focusing on fluency, flexibility, originality or elaboration (see Table 5.2 for a model of thinking developed from Moseley et al. (2005) which indicates the different areas of thinking which can be managed). If you can't identify the kind of thinking you are being metacognitive about, you are not being metacognitive.

---

[2] This contemporary take on a metacognitive recall strategy comes from Ancient Greece and the poet Simonides of Ceos (reputedly). It is known historically as the 'method of *loci*' (places). Simonides was asked to identify the remains of banqueters after the collapse of the hall. He then named each body based on visualising where they had been seated in the hall.

Table 5.2: *Classifying thinking*

STRATEGIC AND REFLECTIVE THINKING				
METACOGNITION Engagement with and management of thinking/learning				
SELF-REGULATION Engagement with and management of values, feelings and motivation				
PSYCHOMOTOR	**COGNITION**			**AFFECT AND MOTIVATION**
	Information-gathering	*Building understanding*	*Productive thinking*	*Affective thinking* *(feelings and values)*	*Motivational thinking* *(conation)*
Consciously directed movement and physical action	Experiencing, recognising and recalling Comprehending messages and recorded information	Development of meaning (e.g., by elaborating, representing or sharing ideas) Working with patterns and rules Concept formation Organizing ideas	Reasoning Understanding causal relationships Systematic enquiry Problem-solving Creative thinking	Receiving Responding Valuing Organizing Characterizing	Aspects of mental processes directed by change and including impulse, desire, volition and striving

The potential impact of these approaches is very high but can be difficult to achieve as they require pupils to take greater responsibility for their learning and develop their understanding of what is required to succeed. The level of challenge for the task is central to this, combined with a level of pedagogic trust between teacher and student (Duffin and Simpson, 2000): too easy and there is no need to be metacognitive, too hard and the learner doesn't have the capacity to be metacognitive. The teacher takes into account the level of challenge in relation to their knowledge of the learner. The greater the level of trust on both sides, the greater the challenge that can be offered and accepted. There is no simple method or trick for this. It is possible to support pupils' work too much so that they do not learn to monitor and manage their own learning but come to rely on the prompts and support from the teacher. Setting too low a level of challenge undermines your belief in a learner's capability. 'Scaffolding' can provide a useful metaphor. A teacher might provide additional guidance through prompts and resources when first introducing a pupil to a concept or to new skills then reduce this guidance to ensure that the pupil continues to manage their learning autonomously and increase their capability to work independently. These strategies are often more effective when taught in collaborative groups so learners can support each other and make their thinking explicit through discussion. Sharing a challenging task may also help manage the complexity of the challenge. It is also important to remember the nature of the scaffolding metaphor. When building, the scaffold does not touch the building and is designed to allow people and resources easy access to the level a structure needs to be worked on effectively. I think of it as temporarily reducing cognitive load to help learners develop effective monitoring of the task such as by practising to develop fluency or learning where to direct attention as they undertake a complex task. Unfortunately, the kind of scaffolding you often see in schools is more like a flying buttress (Stone, 1998), which props the learner up in the short term, only to have their capability collapse without this support.

A number of meta-analyses of metacognition and self-regulation over the last 30 years have consistently found similar, relatively high, levels of impact for strategies related to metacognition and self-regulation (see Figure 5.4; further details about these meta-analyses can be found in Appendix A.5.2). Most studies have looked at the impact on literacy or mathematics, though there is some evidence from other subject areas like science, suggesting that the approach is likely to be more widely applicable.

Table 5.3: *Metacognition and self-regulation meta-analyses*

Meta-analysis	Focus	Number of studies	Pooled effect
Losinski et al. (2014)	Self-regulated strategy development for SEN	16	0.90
Haller et al. (1988)	Metacognition for reading comprehension	20	0.71
Klauer and Phye (2008)	Inductive reasoning	17	0.69
Chiu (1998)	Metacognitive reading interventions	43	0.67
Donker et al. (2014)	Learning strategies	58	0.66
Dignath et al. (2008)	Self-regulation	48	0.62
Higgins et al. (2005)	Thinking skills	19	0.62
De Boer et al. (2014)	Learning strategy instruction	58	0.57
Fauzan (2003)	Metacognitive strategies for reading comprehension	24	0.50
Zheng (2016)	Self-regulation prompts in computer-based learning	29	0.44
Abrami et al. (2008)	Critical thinking	117	0.34
Total studies and overall weighted mean effect size		***449***	***0.58***

Digital Technologies

Over the last 40 years the use of digital technologies in education has been researched extensively (see Table 5.4; further details about these meta-analyses can be found in Appendix A.5.3). As each new technology emerges, people speculate about how it might improve learning and then experiment to find out. Approaches in this area are therefore greatly varied. Overall, studies consistently find that digital technology is associated with moderate learning gains (on average, an additional four months' progress per class, or for every 100 learners experiencing the approach about nine will benefit compared with business as usual).

Overall the evidence suggests that technology should be used to supplement other teaching, rather than replace more traditional approaches. It is unlikely that particular technologies bring about changes in learning directly, but different technologies have the potential to enable changes in teaching and learning interactions, such as by providing more effective feedback, for example, or by enabling more helpful representations to be used or simply by motivating students to practise more and to learn for a longer length of time.

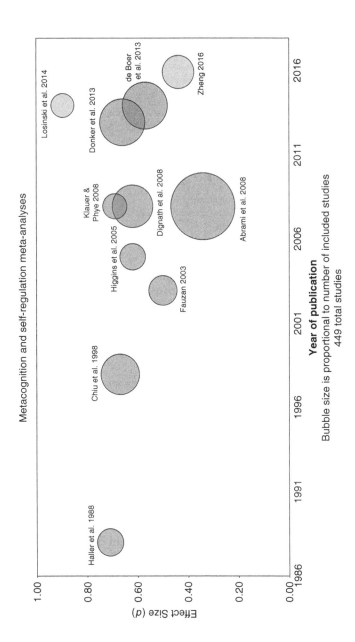

Figure 5.4: Bubble plot of metacognition and self-regulation meta-analyses

Table 5.4: *Digital technology meta-analyses*

Study	Focus	Number of studies (k)	Pooled effect
Torgerson and Zhu (2003)	Writing	2	0.89
D'Angelo et al. (2014)	Simulations	122	0.62
Chauhan (2017)	Elementary education	27	0.55
Morphy and Graham (2012)	Writing	30	0.52
Moran et al. (2008)	Beginning reading	7	0.49
Pearson et al. (2005)	Reading	30	0.49
Tingir et al. (2017)	Mobile devices	14	0.48
Rosen and Salomon (2007)	Constructivist learning environments	32	0.46
Kulik and Fletcher (2016)	Intelligent tutoring	23	0.44
Seo and Bryant (2009)	Mathematics for SEN	11	0.37
Torgerson and Elbourne (2002)	Spelling	7	0.37
Clark et al. (2016)	Game-based learning	55	0.35
Tokpah (2009)	Pre-college algebra	31	0.35
Wouters et al. (2013)	Serious games	77	0.29
Li and Ma (2010)	Mathematics	46	0.28
Torgerson and Zhu (2003)	Reading	4	0.28
Bayraktar (2000)	Science	108	0.27
Onuoha (2007)	Digital labs in science	35	0.26
Kunkel (2015)	Reading	61	0.21
Torgerson and Zhu (2003)	Spelling	4	0.20
Blok et al. (2002)	Early reading	42	0.19
Cheung and Slavin (2012)	Reading	84	0.16
Means et al. (2009)	Online learning	46	0.16
Zheng et al. (2016)	One to one laptops	10	0.16
Cheung and Slavin (2013)	Mathematics	85	0.15
Lou et al. (2001)	Group collaboration	100	0.15
Steenbergen-Hu and Cooper (2013)	Intelligent tutoring systems	26	0.09
Strong et al. (2011)	Fast ForWord	6	0.08
Weighted mean		**1,125**	**0.28**

There is some evidence that it is more effective with younger learners, and studies suggest that individualising learning with technology (such as one-to-one laptop provision or individual use of drill and practice) may not be as helpful as small group learning or collaborative use of technology. There is clear evidence that it has greater impact on areas like writing compared with spelling and mathematics practice rather than problem-solving.

There is extensive evidence across age groups and for most areas of the curriculum that shows positive effects on learning (see Table 5.4).

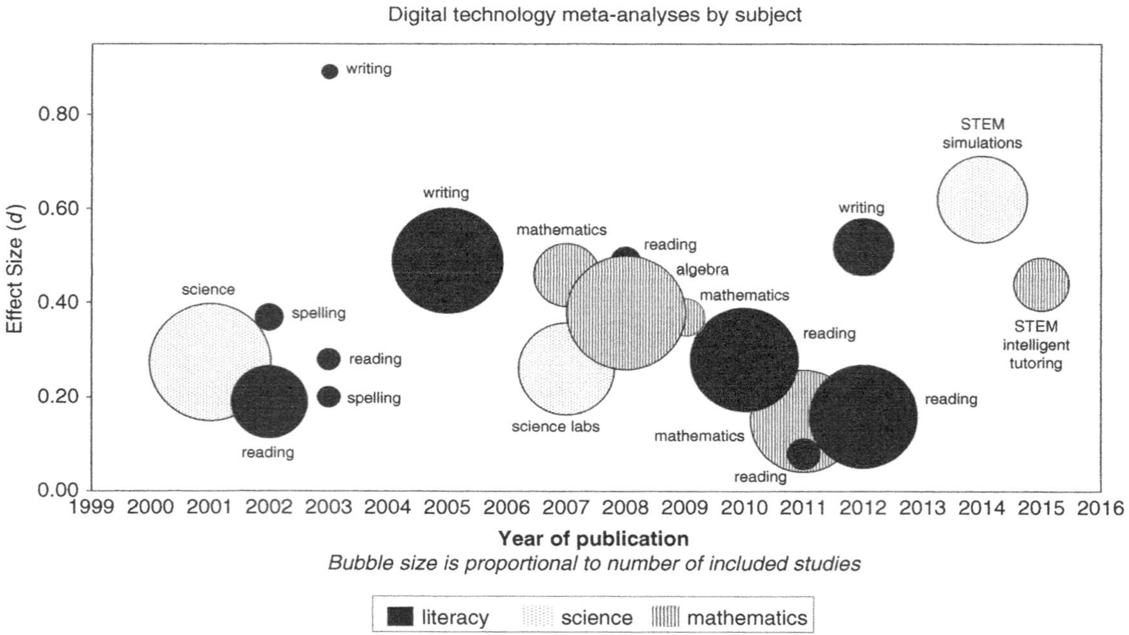

Digital technology meta-analyses by subject

Bubble size is proportional to number of included studies

legend: literacy science mathematics

Figure 5.5: Bubble plot of digital technology meta-analyses by subject outcomes

However, the variation in effects and the range of technologies available suggest that it is always important to evaluate the impact on learning when technology is used. The pace of technological change means that the evidence is usually about yesterday's technology rather than today's. Average effects have remained consistent for some time, implying that general messages are likely to remain relevant.

How Come Everything 'Works'?

It is hard to know if these positive effects say more about the default state of teaching and learning or are a clear message about digital technology. Given that, on average, most things tend to bring about improvement, I wonder if we should see this at least partly as an indication that most areas of teaching and learning can be improved with effort and focus, so it becomes more of a question about efficiency rather than effectiveness. There are all kinds of ways to teach different skills, knowledge and understanding. The question is which are the most effective, and what are the strengths and weaknesses of different approaches? My personal view is that, when you introduce technology into a teaching and learning setting as a teacher, you tend to think more carefully about the teaching and learning activities, as you plan how the technology can be used. The objectives, activities and actions of the learners have to be thought through in detail and aligned carefully. Also, as technology is introduced, it is novel for learners and provides variety in what they experience, so this tends to make them more positive. The more successful attempts (those which are above average in terms of impact) ensure that learners actually do work more efficiently or more effectively (or simply for longer). The technology helps you put some aspects of teaching and learning under the microscope and opens up or unfreezes the interaction and experiences of the teacher and the learners. Over time, these interactions settle down and refreeze into more routine patterns. Sometimes these are better than before, sometimes worse, often much the same. The subject patterns (see Figure 5.5) seem to indicate that writing is one area which has potential for improvement. This is a difficult area to teach successfully, not least because becoming an effective writer requires extensive practice. It is therefore more likely to be amenable to improvement. Reading is probably one of the most researched fields in education and overall reasonably well taught. There is therefore less gain to be found in new approaches, compared with what we usually do. The room for improvement is smaller.

My final concern about digital technology is that people are always ready to attribute improvement to the technology itself. This is evident in the focus of many of the meta-analyses on a particular technology. This concerns me because the technology does not have agency. If there is improvement, this must result from what the learners do differently. They still have to spend time and extend effort in their learning. We can't insert a chip or download new skills, knowledge or understanding into the human brain. Our knowledge and skills are interconnected and integrated with our existing knowledge and capabilities. Adding 'storage' of new knowledge is problematic as this would need to be integrated or experienced in some way by the recipient. This is also why some aspects of school learning are soon forgotten as they are like modular plug-ins; they are only referred to and connected with other isolated school experiences. Once they have been rehearsed for assessment, they are no longer needed or accessed. Technology is a tool which can help teachers teach more effectively or a tool for learners which makes it easier for them to engage with the content or receive more feedback. On other occasions technology might be beneficial because it makes aspects of learning more difficult but in a way which encourages effort and success. A blackboard, a book and a pencil are older forms of educational technology, but they are not seen as having agency. It is valuable to reflect on whether digital technology is really any different from these earlier types.

Learning Styles: A Pedagogical Hydra

The Toolkit has an entry on learning styles, and this has received some criticism for a number of reasons, particularly that the estimate of impact is + 2 months, which appears to support the use of such approaches. Although such techniques such as VAK (visual auditory kinaesthetic) were very popular about 10 or 15 years ago (and were even endorsed in official documentation and by Ofsted, England's school inspection organisation), there is now a general consensus that such ideas are not psychologically or scientifically robust. My understanding of the current state of knowledge in relation to learning styles is that, as individuals, we don't really have a preferred 'style'; our preferences tend to be task specific, and even when we think we prefer an approach we actually choose to undertake tasks differently if observed (Pashler et al., 2008). Furthermore, when such styles are identified, the instruments used are unreliable, and if you try to target or match tasks with such preferences it does not lead to improved outcomes, at least as indicated by the most robust studies (Cuevas, 2015). In short, 'learning style' is not a robust psychological concept. The confusion about learning styles persists, however, even in the psychological literature. Komarraju and colleagues'

Table 5.5: *'Learning styles' meta-analyses*

Meta-analysis	Focus	Number of studies (k)	Pooled effect
Kavale and Forness (1987)	Perceptual learning styles	39	0.14
Garlinger and Frank (1986)	Matched cognitive style	7	0.03
Lovelace (2005)	Dunn and Dunn model	76	0.67
Tamir (1985)	In science	18	0.06
Slemmer (2002)	Technology enhanced learning	48	0.13
Kanadli (2016)	In Turkey	29	1.03
Total studies and overall weighted mean effect size		***217***	***0.16***

study in *Personality and Individual Differences* in 2011 (recognised as a top journal in social psychology) identified 'the four' learning styles (apparently these are synthesis/analysis, methodical study, fact retention and elaborative processing) and found they explained only 3 % of the variance in grade point average of undergraduates (i.e., in relatively successful learners) over and above personality differences. The (very small) correlations indicated that the most relevant dimensions were the reflective 'learning styles'. I can't help wondering whether self-regulation and metacognition might not have been better constructs for the authors to use, as it appeared to be strategic decisions to adopt different 'styles' and learners' conscientiousness, which accounted for the approaches of successful learners.

Overall there is an academic consensus that learning styles are not a valid idea, but what makes the concept so appealing? The first studies I am aware of in this arena looked at visual and auditory approaches to teaching reading in the 1950s and 1960s. It wasn't particularly successful then either (see Higgins, 2013). The idea resurfaces perennially and is hard to eradicate, like a modern pedagogical hydra. I think a number of features sustain the notion in the educational ecology. A key issue is the idea that you might be able to meet learners' needs better if you can take their individual capabilities into account. This drives a range of ideas in education such as differentiation, ability grouping and personalisation. I believe this results from the demand of trying to address the needs of large groups of students with varying capabilities as a sole teacher. I worked with a number of teachers in a 'learning to learn' project (Higgins et al., 2007) in which teachers in the early phases of the project were convinced of the usefulness of learning styles approaches. As we introduced more rigorous evaluation techniques, most of those involved chose to focus on other areas like metacognition or specific approaches to develop learners' resilience or subject-specific strategies. When subject to sustained professional enquiry, other ideas proved more useful. But it is helpful, I think, to consider what happened in practice in classrooms when teachers adopted 'learning styles' techniques. I am personally convinced that one potential benefit derived from the teacher's move in providing choice to learners. Some teachers, under the banner of 'learning styles', planned a range of activities which they believed catered to different learning 'styles' but which met the same lesson objectives. They then asked learners to choose which they thought fitted their 'style'. When this worked, there was a subtle shift in responsibility where the learners engaged with the task differently as a result of the proffered choice. There is another change here too where teachers' planning was undoubtedly invigorated, they took great care to plan a range of ways to teach specific skills and knowledge. On the other hand, there were also some examples where the approach was potentially destructive and where teachers classified learners as, for example, kinaesthetic and then assumed they were or should be limited to this style. This even extended to affecting some learners' self-belief. I interviewed some children as part of the project. One of them, let's call him Rahil, said 'I can't write. I'm a kinaesthetic learner'.

I am certainly not advocating the use of learning styles approaches. Inaccurate and unreliable labelling of pupils is too great a risk. However, I do get frustrated with critics who become dogmatic that learning styles do not exist, so nothing associated with their practice can ever be effective in the classroom (with the implication that foolish teachers were duped by the rhetoric into harming their pupils). The teachers I worked with were dedicated and committed, but they perhaps developed activities which were effective for other reasons or interpreted learning styles in a way which mitigated some of the risks. There was also some confirmation bias too, reinforcing what was a plausible, but inaccurate, belief. I think of this as a false form of catalytic validity (Lather, 1986) in which the changes in practice and feedback from learners were genuine, and these drove positive changes in the classroom, even though the underpinning theory and constructs were incorrect. It was inefficient, therefore, rather than necessarily damaging. Most teachers work extremely hard to mitigate any potential damage entailed by policy prescriptions. However, the idea also prevented teachers from identifying any beneficial causal mechanism by invoking a misleading pupil characteristic. This was the main issue for me. 'Learning styles' was seen as a solution which blocked further enquiry and experimentation into how you balance individual and group needs in a class, which always will be an enduring challenge in education, especially with the current resource allocation.

The evidence from research fits this interpretation; overall it is weakly positive, but somewhat inconsistent (see Table 5.7 and Figure 5.7, with further details about these meta-analyses in Appendix A.5.4). The most recent meta-analysis (Kanadli, 2016) is astonishingly positive (d = 1.03). It is composed of many small studies, mainly masters and doctoral theses conducted in Turkey investigating the effect of instructional designs based on learning styles models. Most were quasi-experiments with only four true experiments (presumably these included randomisation, though this is not explicit), these four studies showed an effect size of 0.65, substantially lower but still not similar to earlier meta-analyses where the effects range from 0.03 to 0.13). The Lovelace (2005) meta-analysis has been criticised in the literature (Kavale and Le Fever, 2007) for similar reasons: 35 out of 36 included studies are dissertations by students of the proponents of the particular learning styles inventory (the Dunn and Dunn model). Both appear broadly technically sound, though concerns about the robustness of the underlying concepts must increase doubt about the findings and implications.

I am reminded of Pratt and Rhine's ESP review from the 1940s (see Chapter 2). The Kanadli, (2016) and Lovelace (2005) meta-analyses appear to be robust and genuine attempts to evaluate the impact of learning styles approaches but are not consistent with other meta-analyses (see Figure 5.6). However, it seems likely that there are weaknesses in the underlying designs of the included studies and in the frame that the authors set to identify studies. My guess is that the designs did not evaluate the impact of learning styles accurately or precisely enough and that the positive results are where other changes were responsible for the benefit (such as in teacher planning or pupil engagement and choice). This is not unlike my analysis of the questionable value of introducing new digital technologies. When there are positive effects we look only at the technology, not the other changes introduced or entailed which may be as responsible for any improvement. The key difference is that the digital technologies have substance, whereas the impact of learning styles is delusive.

The research evidence on learning styles provides a challenge to meta-synthesis. Should it be included, as it is an approach which is periodically adopted by teachers? Or should it be excluded because the underlying concepts are problematic? On balance, I argued it was best to include the entry, warts and all, as this was most consistent with our overall rationale and methodology. Although I have concerns about the

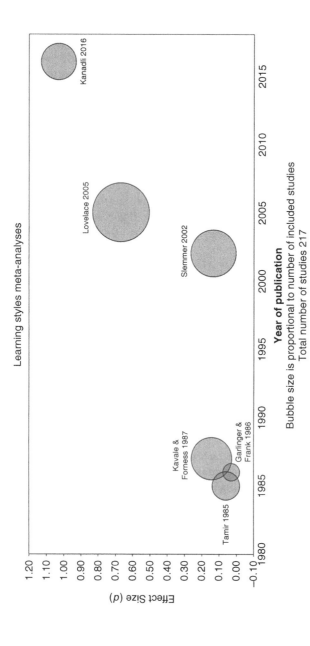

Figure 5.6: Bubble plot of learning styles meta-analyses by year

underlying quality of the evidence, we can't pick and choose what this says. The entry on learning styles is clear and states that we think the impact results from other changes, due to the lack of conceptual clarity about what learning styles are and due to problems in reliably assessing them (Coffield et al., 2004). This can be easily misinterpreted at first glance.

Chapter Summary

This chapter has provided an overview of some of the comparative findings between meta-analyses about pedagogy and some reflections about some of the more effective and some of the less effective approaches (on average) to teaching and learning. One broad inference is that the likelihood of impact decreases the further you get from teaching and learning interactions. Areas like feedback or metacognition and self-regulation appear to have the most positive effects, on average, so offer a 'good bet' for teachers and schools. Structural changes such as grouping and timetabling make less difference, on average, so schools would need to be really clear about what they think the causal pathway to learning is, if the aim is to improve outcomes for learners. Most other people have failed to have an impact with structural approaches. Why do you think you could be different? In my view, they are high risk investments of time and money, so schools need to be clear about why they think they can buck the trend. They would need to be substantially better than average to have a positive effect. This idea of a causal pathway is relevant to other areas such as the use of digital technologies or learning styles. In these areas, we need to be clear what actually happens in practice in terms of what learners do. For learning styles, the concept is not sufficiently coherent to be helpful, and the idea may encourage teachers to focus on the wrong aspects of the teaching and learning process, though some changes implemented based on the idea may still be effective. Ideally, we want to bring about improvement and to achieve this for the right reasons as this will help focus on what is really making the difference. The challenge with technology is that we often focus on the technology and the activity, so it is easy to lose the focus on the learning.

What Don't We Know

We don't know to what extent some of the patterns in effects are related to aspects of the research process or issues with the underlying research. It is not surprising that feedback interventions tend to have the highest effects because they aim to directly improve performance. However, feedback is a complex idea, and the evidence is not sufficiently coherent or detailed to provide specific suggestions about the most effective ways to provide feedback in the hectic classroom environment. For metacognition and self-regulation, again the research suggests that this is a promising area which supports better outcomes for learners, but again the specific detail about what, when and for whom is lacking, beyond general inferences for older learners and the benefits for low attainers. The contrast between the research on digital technologies and learning styles is instructive, and we need to be consistent in the conclusions we draw, rather than be influenced by our beliefs and assumptions about what we think should and should not be effective.

Final Thoughts

For the opening to this chapter I used a quotation from 'T'aint what you do, (it's the way that cha do it)', written by the jazz musicians Melvin 'Sy' Oliver and James 'Trummy' Young in 1939. It was made famous by Ella Fitzgerald and the 'Shim Sham' tap dance routine. Most people today are more familiar with familiar with the Ska beat of the 1982 Bananarama and Fun Boy Three version of the song, and this makes a nice humorous aside in presentations. I've called it the 'Bananarama principle' as a way of remembering the key idea that when using evidence from research 'it ain't what you do, it's the way that you do it'.

The underlying point is a serious one, and I see it is a way of expressing the more challenging statistical idea that the variation within each Toolkit strand is greater than the variation between strands. It is important to focus on the spread of effects as well as the average. With feedback, the mean effects are high when you look across the available meta-analyses and compare these with other educational interventions. However, there are also some studies which show the negative effects of feedback on performance and learning. Although the average gives you a 'good bet', in that you would expect a similar pattern of effects over time if people try similar approaches, there is also a risk that you might be more similar to the unsuccessful studies and you could make things worse. In the original version of the song, Ella Fitzgerald sings 'It ain't what you do, it's the *time* that you do it, that's what gets results'. This is particularly true for feedback as the timing of the information is crucial: too early and you risk prompting the learner unnecessarily, too late and the moment has passed.

6   Meta-analysis and Literacy

Key questions
What can we learn about the teaching of literacy from
 meta-analysis?
Can any of the key debates about reading be settled?
What can we learn about meta-analysis from research into
 literacy?

The Dewey Decimal System consisted, in part, of Miss Caroline waving cards at us on which were printed `the', `cat', 'rat', 'man', and 'you'. No comment seemed to be expected of us, and the class received these impressionistic revelations in silence. I was bored, so I began a letter to Dill. Miss Caroline caught me writing and told me to tell my father to stop teaching me. 'Besides,' she said, 'we don't write in the first grade, we print. You won't learn to write until you're in the third grade.'

Harper Lee *To Kill a Mockingbird*, 1960

Introduction

This chapter reviews the evidence from meta-analysis on four areas of literacy interventions: early years, speaking and listening (oracy), reading and then writing. It summarises my understanding of the overall evidence and identifies some of what I see as the specific issues in each area in using evidence from meta-analysis. Some of the implications for teaching and policy are then drawn out.

Literacy is perhaps the most researched area of schooling, particularly early literacy development in young children, together with beginning reading and the further development of reading capability. However, as we saw in Chapter 1 when looking at phonics, there is still much that we do not know. We also have to be careful by what we understand as the implications from the experimental literature, as, from my viewpoint, these need to be understood through the lens of the methods used.

Evidence from Early Literacy Interventions

Approaches which aim to improve the skills, knowledge and understanding of young children in relation to reading or writing are extensive and diverse. Common themes include book sharing, storytelling and group reading, as well as activities that aim to develop letter knowledge, knowledge of sounds and early phonics, as well as introductions to different kinds of writing. Other elements are often included, such as involving parents (see Chapter 7) or an emphasis on developing speaking and listening skills. The diversity of these approaches reflects the ambiguity in the adjective 'literate', which can mean both able to read with a focus on capability and skills but also more widely 'educated' in terms of culturally valued knowledge and understanding.

Early intervention approaches have been consistently found to have a positive effect on early learning outcomes when researched experimentally (see Figure 6.1 and Appendix A.1.1) for more details about these meta-analyses). All children appear to benefit from early literacy approaches, but there is some evidence that certain strategies, especially those involving targeted small group interaction, may have particularly positive effects on children from disadvantaged backgrounds. Early literacy approaches are not a panacea, however. Though long-term positive effects have been detected in some studies, these are much more elusive, and for a majority of approaches these benefits appear to fade over time. As mentioned above in Chapter 5 in the section on treatment inherent and treatment independent measures, about half of the gain is lost in the first year, then the fade flattens out to an average effect size of about 0.01 per year. Interestingly this analysis (Li et al., 2017/in preparation) also indicates that cognitive gains (such as verbal comprehension and non-verbal reasoning) appear more resilient than academic measures (such as letter knowledge).

There is also evidence that a combination of early literacy approaches is likely to be more effective and resilient than any single approach. For example, some studies suggest that it is possible to develop certain aspects of literacy, such as knowledge of the alphabet or letter names and sounds, without improving other aspects of early literacy. Taking this with the evidence about the greater decline in effects from academic measures, it is likely to be beneficial to put a range of activities in place and to use these in combination with diagnostic assessments of early literacy skills, knowledge and understanding. The bubble plot (Figure 6.1) suggests that the average effect has remained consistent over time but that more recent research has attempted to separate out some of the components of effective early education programmes. Here it is difficult to know whether

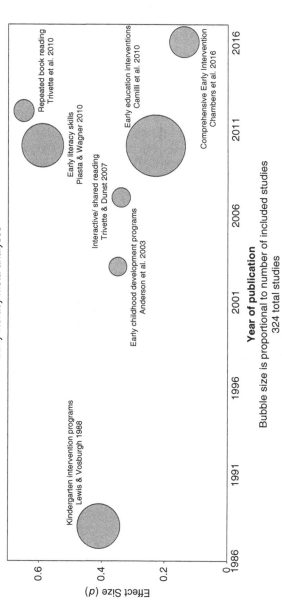

Figure 6.1: Bubble plot of early literacy interventions

some approaches are generally more effective or whether some were better targeted for a particular intervention group and were successful because of this.

The 'Proxy Problem'

One of the many challenges in effective assessment and testing is what I think of as the 'proxy problem'. This is where an indicator of future literacy achievement, such as early letter knowledge or the skill of writing your name, is then used as a target for literacy teaching. We don't teach letters and sounds as an end in themselves, but they are a necessary step in learning to read. The problem with this is that, when some tests were initially devised, these indicators stood as proxies for a wider range of aspects of growing literacy capability. They were selected as assessment points because they could be assessed reasonably reliably and because they correlated well with other developing skills and knowledge as a predictor of later success. These assessment data points provided validity because they identified a particular skill or aspect of knowledge which mapped across other developing skills and knowledge and were a good indicator of subsequent success. I think of the measure as sweeping up a range of other skills and knowledge but acting as an implicit indicator of them. However, when assessment becomes more about the teacher or the school than the individual child, the validity of these assessment points is compromised as teachers' efforts focus on success at the assessment task, rather than the wider underpinning skills and knowledge. An example of this is the non-word screening test in England. The capability to read non-words (such as splog, or gep[1]) correlates reasonably well with real word reading (about 0.85). A number of assessments were developed in the 1990s to help with the identification of reading difficulties. As a diagnostic activity, this can be very useful. It helps quickly identify children's decoding capability. However, once it is used as a high-stakes assessment, it becomes used to evaluate school performance. This creates a backwash[2] where children are deliberately taught to read non-words as a performance skill. This strikes me as somewhat risky. Of course, we

---

[1] This should be pronounced with a soft 'g', so 'jep', on the basis that about 64% of words starting with 'ge-' in English have a soft sound. I find this problematic as the most similar word in English is 'get'. I also went to some lengths to find out how to pronounce Ursula Le Guin's character 'Ged' correctly in 'The Wizard of Earthsea'. According to the author, it should be a hard 'g'. In a non-word reading test both pronunciations should be correct, but this is not always the case.

[2] Stephen Wiseman commented on the backwash effects of examinations in 1961 in the context of the 11 plus examinations in England, observing that little effort had been expended in mitigating these effects (Wiseman, 1961: p. 158).

need children to decode words fluently and rapidly. Mastering the sound/ symbol code in English is more challenging than in many languages. Nearly all children will need some explicit teaching of letters and sounds at some stage. However, we don't want to stop children reading for meaning and using context to determine the word and its correct pronunciation where appropriate. Here I'm thinking of words like 'lead' (for a dog) and 'lead' (weight) or 'bow' (of a ship) or 'bow' (tie). The challenge is to make sure that the emphasis between decoding and comprehension shifts appropriately during teaching to ensure efficient progress (see the section on learning to read and reading to learn later in this chapter for a further discussion of this balance). For interventions in the early years which aim to develop literacy, we need to be clear which genuinely contribute to more rapid progress over the medium to long term and are not just examples of 'squeezing the balloon' (see Chapter 5) to accelerate one aspect of development, only to find that long-term growth and achievement are not improved. My worry is not just that some approaches which focus on a narrow set of skills may be ineffective, as the gains are more likely to fade, but we also have to be concerned about what else the children could have been learning, or the educational opportunity cost, to use an economics metaphor.

At this time, there are many things we still don't know for certain, but the evidence indicates that a number of early years interventions to improve literacy outcomes can be successful. The 'best bets' from meta-analysis for early years intervention would be to start young (certainly by age 3) to keep a focus on thinking and learning (rather than specific sets of discrete academic skills) using a broad programme with an integrated set of activities and goals. Then it will be important to ensure that subsequent teaching and learning aims to build on earlier accomplishments to try to reduce the fade-out of early benefits. Diagnostic assessment which helps to plan future activities and support for individual children or small groups should be encouraged, but using the same assessments as benchmark targets for cohorts and schools should be avoided or at least kept separate from everyday practice to avoid jeopardising the validity of early assessment measures.

The Importance of Developing Spoken Communication for Learning

Speaking and listening approaches and interventions emphasise the importance of spoken language and verbal interaction in the classroom. Whilst effective communication capabilities are an important outcome from education, the justification for their explicit inclusion in schooling is usually an instrumental one in that they help develop reading comprehension and writing outcomes. Here the idea is that that both reading comprehension and writing skills can benefit from explicit discussion of either the content or processes of learning, or both. Speaking and listening approaches language approaches typically include:

• reading to and discussing books with young children (see above);
• explicitly extending pupils' spoken vocabulary;
• the use of structured questioning to develop reading comprehension;
• the use of talk to develop articulation of content for writing and to increase self-regulation of the writing process (such as to revise and improve or self-talk to persist and succeed).

Some aspects of speaking and listening interventions overlap with a number of other approaches. This includes approaches based on metacognition and self-regulation, which make talk about learning and reasoning more explicit in classrooms, and with collaborative learning approaches, which promote pupils' talk and interaction in groups. Similarly, peer tutoring and reciprocal teaching approaches involve structured spoken interaction, as do a number of techniques which teach strategies for reading comprehension.

I should be clear about my own position here, in that I have always been concerned about the lack of emphasis on spoken language and interaction in our assessment system in the UK. This may reflect a personal bias, in that I felt I benefited from learning to debate formally at school. I have always been interested in the history of rhetoric and the dangerous art of persuasion and argument. It also results from my experience of teaching in schools in disadvantaged areas across North-East England. I became aware of the disadvantages of restricted codes in an educational system which understandably privileges more elaborated codes. These codes also depend on an increasing precision in vocabulary, drawing on Latin and then ancient Greek vocabularies for this precision and detail.

These ideas come from Basil Bernstein's work identifying differences in language through class and identity. A restricted code is based on close social relationships with a:

common, extensive set of closely-shared identifications and expectations self-consciously held by the members. The speech is here refracted through a common cultural identity which reduces the need to verbalise intent so that it becomes explicit ... The speech in these social relations is likely to be fast and fluent, articulatory clues are reduced; some meanings are likely to be dislocated, condensed and local; there will be a low level of vocabulary and syntactic selection; the unique meaning of the individuals is likely to be implicit. (Bernstein, 1971: pp. 125–126)

By contrast an elaborated code is much less predictable and:

is likely to arise in a social relationship which raises the tension in its members to select from their linguistic resources a verbal arrangement which closely fits specific referents ... The preparation and delivery of relatively explicit meaning is the major function of this code. The code will facilitate the verbal transmission and elaboration of the individual's unique experience ... (Bernstein, 1971: p. 127)

So, the language of the classroom requires more knowledge of the content of the conversation compared with the language of home which needs greater understanding of the immediate context. School language is more precise and explicit, home language more implicit and inferential. This has profound implications for what and how we teach. I find it helpful to think of each subject as its own language, or at least a dialect. Practice in speaking these dialects helps learners to negotiate the elaborated code of the classroom.

I am also influenced by theories of language acquisition and the emphasis on both reception and production in language learning. We ask children and young people to listen, rather than speak, most of the time in school. However, if we think of curriculum subjects as being like dialects of an elaborated academic code (with their own specific vocabularies and reasoning rules), then learners need to practice speaking these dialects to learn the subject thoroughly. Otherwise they will be fluent in comprehension but not able to articulate or write subject content. I can read and comprehend written French tolerably well. I can hardly speak a complete sentence and certainly can't write one accurately. The same can be said of a typical school pedagogy where young people learn to listen and read geography or chemistry but have less opportunity to be fluent speakers. For those more exposed to an elaborated code outside of school, as Bernstein (1971) argued was the case for 'middle-class' families, young people only have to learn the vocabulary and particular reasoning rules so can make this translation from spoken and written input to written output. For those who are more familiar with a restricted code, they have to learn a new grammar and syntax as well as vocabulary and reasoning when speaking geography or mathematics (Higgins, 2003). A typical school pedagogy omits the spoken stage and expects the learner to go directly from input to fluent written output. For those less successful and less fluent in the subject this can be too big a step. Just to be clear, I see this explanation as a metaphor to help understand children's difficulties in academic language. I am not advocating lessons in practicing to speak geography, for example. However,

I think the metaphor can be a useful one in seeing why some children and young people find it difficult to articulate their ideas and to write fluently in different lessons and subjects.

Speaking and listening activities can support pupils in expressing their current understanding in a way which practises specific skills and content for writing. When writing, learners need to think about purpose and audience, as well as the coordination of meaning, form and structure. This coordination is a complex skill that can be practised through purposeful speaking and listening activities as preparation for the content and structure of a piece of writing. In the classroom, such activities are also an assessment opportunity for the teacher to understand pupils' fluency with the particular grammar, syntax and vocabulary needed for a piece of writing.

Speaking and listening has been the subject of a number of meta-analyses (Figure 6.2). What makes interpreting these challenging is that the focus varies across different areas of literacy. Reading has unsurprisingly been of perennial interest but includes reading to, reading with and reading aloud, all with similar levels of impact. Some areas also have apparently contradictory results such as vocabulary instruction (Marulis and Neuman, 2010; Elleman et al., 2009) though here the general conclusion is that teaching vocabulary can be successful and is an important component of reading comprehension but may not be sufficient to close the gap between disadvantaged learners and their more advantaged peers.

Learning to Read and Reading to Learn

Two key aspects of reading have been extensively explored through experiments: phonics and reading comprehension approaches. These strands relate to differing but complementary perspectives about what is important in learning to read. As usual when there is an apparent dichotomy, the answer is a bit of both[3]. The questions we have yet to answer are how much of an emphasis on phonics should there be and how much on comprehension, and when?

---

[3] I was lucky enough to be taught basic quantitative methods by Professor Carol Fitz-Gibbon when she worked at Newcastle University. Her course had a fearsome reputation for its level of difficulty, and she was somewhat eccentric with what were considered to be maverick ideas in education in the 1980s: meta-analysis, randomised trials, evidence and value-added. She had a number of memorable sayings such as 'Correlation, eh? It will all end in tears!' and 'It's never A or B, the question is always how *much* of A and how *much* of B.'

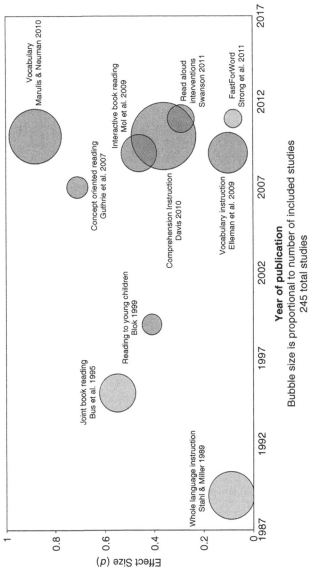

Figure 6.2: Bubble plot of speaking and listening meta-analyses

Phonics

There is extensive evidence that the teaching of phonics can help some readers of all ages in learning to read. The purpose of phonics is to develop one aspect of pupils' phonemic awareness, which is their ability to hear, identify and use phonemes (the smallest unit of spoken language) and to teach them the relationship between phonemes and the graphemes that represent them (letters or combination of letters used to represent a phoneme). Phonemic awareness is a subset of phonological development where young children become capable of discriminating between smaller and smaller units of sound in language as the literacy abilities develop. In practice, many phonics programmes contain elements of wider phonological awareness and capability, such as tapping out the number of syllables in a word or recognising rhyme and alliteration.

Rapid and fluent decoding of words is an essential skill for reading, and the experimental trials show that phonics approaches can be successful for different ages and particularly for those who are not making expected progress. Overall the picture from experimental trials has been fairly consistent for the last 20 years or so (see Figure 6.3, with more details about the meta-analyses in Appendix A.1.1).

Reading for Understanding

Another area of research which provides a contrasting perspective on reading is the experimental evidence from approaches which focus on comprehension. Again, the evidence is extensive and positive (see Figure 6.4, with more details about the meta-analyses in Appendix A.6.1). When researchers have tried to improve outcomes in reading using approaches which focus on understanding of text they have typically been successful. A wide range of strategies and approaches have been developed, but they all focus on the importance of deriving meaning from a text and understanding what is written as the main goal.

Successful readers monitor their understanding of what they are reading and review the text when something does not make sense. Reading comprehension can be improved by teaching specific strategies that learners can apply to both check how well they understand what they read and overcome the difficulties to comprehension that they face. These strategies involving focusing attention on specific skills such as prediction, questioning, clarifying and summarising. This is why metacognitive approaches are valuable, as a learner has to take

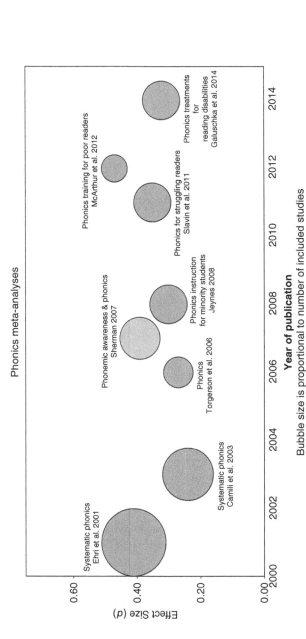

Figure 6.3: Bubble plot of phonics meta-analyses

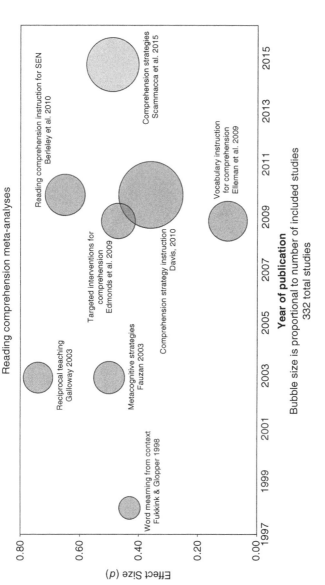

Figure 6.4: Bubble plot of reading comprehension meta-analyses

responsibility for understanding what she or he reads. These strategies can initially be taught and practised separately using carefully selected texts, but learners will also need practice in combining strategies to develop effective comprehension of different texts and of different kinds of text. The effectiveness of teaching readers to integrate multiple strategies is supported by research evidence and is likely to be more effective than relying on single strategies in isolation. Overall, the aim is for the readers themselves to take responsibility for automatically using these strategies to monitor and improve their understanding. Again, this is why metacognitive approaches are so useful (see Chapter 5).

The potential impact of these approaches is high but can be difficult to achieve in terms of targeting appropriate skills in relation to learners' current capabilities and in terms of text selection. The aim is for learners to increase the fluency of their reading skills and to apply the techniques so that they become automatic. Text selection needs to take account of background knowledge. The text needs to be difficult enough to make application of the strategy helpful but not so difficult that they can't apply the strategy successfully. Text difficulty can relate to two main dimensions, structure and content. Usually comprehension strategies focus on text structure; however, the difficulty may lie in the context and content (such as vocabulary), and focusing attention on the grammar or sentence structure may not be of much help.

What the evidence does not tell us is which approach is likely to be most beneficial for a particular learner or class. Both approaches can be valuable, but one of the disadvantages of meta-analysis is that, as we aggregate the effects of interventions, we are more confident that such approaches can be successful but become less clear about precisely what they are better than. I think of this as averaging the control groups, as well as averaging the intervention effects. To enable us to target appropriate approaches for learners, we really need to know more about the comparison groups. A reasonable interpretation would be that once children's phonological development is progressing well, the emphasis on comprehension needs to increase. If difficulties arise with the progress a child is making, it is important to be clear what to focus on. It may be necessary to increase the automaticity of decoding and emphasise this fluency. However, it may also be an issue with knowledge and vocabulary of other aspects of comprehension. Being clear of the cause of the difficulties will enable a solution to be identified more efficiently.

International evidence indicates that we underestimate the difficulties some young people are still having with decoding, even once they are at secondary school (van de Ven et al., 2017). Their skills may not be sufficiently rapid and automatic to be able to think about the meaning of what they read. This is particularly a challenge because they need to be able to read well to be able to learn from they're reading. Typically, in secondary school young people also need the background knowledge and understanding to be able to learn from expository texts across a range of subjects.

I think that the evidence indicates that both approaches, phonics and comprehension, are necessary and jointly sufficient. I am influenced in this by the simple view of reading (see Figure 6.4). It is important to remember that the simple view of reading is just what it claims. It is a simple view which sets two dimensions of reading, decoding and comprehension, as the main dimensions. Reading is, of course, more complex than this, and within each area there is more to be said, whether you start to look at the complexity of meaning through inference and connotation (which may depend on wider cultural knowledge and understanding) or whether you start to consider individual difficulties in phonological processing and what may prevent a learner decoding fluently.

The Simple View of Reading

The Simple View of Reading (Figure 6.5) is best known as a diagrammatic representation of a model of basic reading where reading comprehension (RC) is defined as the product of listening comprehension (LC) and decoding (D). $RC = LC \times D$. It was developed in the 1980s, when the 'reading wars' raged (Gough and Tunmer, 1986: see Chapter 1). It was proposed as way to reconcile the two opposing camps of early literacy teaching – a meaning-led approach on the one hand and phonics on the other.

Gough and Tunmer argued that the Simple View of Reading acknowledges the value of aspects of a meaning-led approach by positioning reading as a linguistic activity but also gives phonics an essential role. The compromise was not accepted by either side, with some arguing that the complexity of linguistic competence and the reader's expectation of text was not recognised, whilst others found insufficient detail and rigour about the processes involved in decoding. In England, it was adopted by the Rose Report and formed a key part of the Primary National Strategy's view of literacy learning (Wyse and Styles, 2007), and it is also used as an illustrative model in the Education Endowment Foundation's guidance on literacy in primary schools (EEF, 2017).

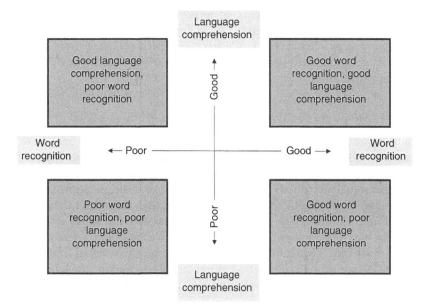

Figure 6.5: The Simple View of Reading

There is a broad consensus supported by research that reading requires both comprehension and decoding, with acknowledgement that the Simple View is only a very basic model. For professionals in the early years, for example, it is important to understand that listening comprehension involves knowledge of language forms (morphology and syntax) as well as vocabulary and understanding of the context (semantics and pragmatics). Many academics consider that the relationship between listening comprehension and decoding is not shown adequately by the Simple View. Overall the evidence supports the interpretation that decoding and listening comprehension are both necessary but that separately each is not sufficient for skilled reading (Savage et al., 2015). As I argued in Chapter 1, we have not developed the evidence base to improve the precision about how to make decisions about this overall balance and how this should shift relative to age or precisely when and how to support learners who are not making expected progress. I believe we need to use better diagnostics routinely to help identify what is most likely to be helpful for a child or class, but we also need more evidence to help us understand what is sufficient at a given stage. This would help teachers to understand better when to stop and move on or when further practice and consolidation is needed. This is important for both phonics and comprehension-led approaches. It is important to know when to stop, as well as when to start. We have extensive evidence from meta-analysis about effectiveness but relatively little about efficiency. When is enough, enough?

Messages for Writing from Recent Meta-analyses

In this section, I've taken a slightly different approach and have summarised what I see as some of the main implications for writing from across a number of meta-analyses. This is partly because most of the meta-analyses on writing do not focus on a single approach or technique to writing to see whether it is effective, but rather they look at a range of studies to draw implications about the most effective approaches. This makes them more similar to a meta-meta-analysis or meta-synthesis. This summary draws heavily on the work of Professor Steve Graham who (in my view) has published some of the most helpful analyses of both typically developing writers and those with non-typical development.

The Range of Effects and the Challenge of the Counterfactual

One of the key issues here is the range of effects. The summary points in Table 6.1 indicate that a number of different approaches and emphases have had success when researched. This is in terms of improving

Table 6.1: *Effect sizes for different aspects for writing*

Summary	Overview	Effect sizes	Reference
Following a structured writing process with clear stages	Pupils are trained to use (and can show in their work) a structured writing process which explicitly includes the stages of planning, drafting, revision and editing. They make use of this process for their writing tasks.	**1.20** **0.82** **0.37 (quality)**	Graham et al. (2012) Graham and Perrin (2007) Graham et al. (2015)
Working collaboratively on writing	Pupils work on their writing in pairs or groups at various stages of the writing process: planning (pre-writing),	**0.89** **0.75**	Graham et al. (2012) Graham and Perrin (2007)

Table 6.1: (*cont.*)

Summary	Overview	Effect sizes	Reference
	drafting, revising, editing to develop or practice skills.	**0.66**	Graham et al. (2015)
Receiving feedback about the quality of their writing which they can act on	Pupils receive regular feedback about the quality of a text they have written from adults, peers or through self-assessment ratings (such as by using a rubric to identify good features). The impact of timely teacher feedback for younger writers is especially large.	**0.80** *adult* **0.37** *peer*	Graham et al. (2012)
Setting own writing goals	At various points in the writing process (planning, drafting, writing, revising), pupils are encouraged to identify specific goals; they later check and report (to the teacher or a peer) whether they have actually achieved those goals. This might include finding at least three sources to use, or adding five supporting details when revising a persuasive text, or drafting an introductory paragraph in a writing session.	**0.76** **0.70**	Graham et al. (2012) Graham and Perrin (2007)
Using digital tools to write	Pupils become fluent in keyboard skills and have regular access to word-processing software when writing which they can use to edit and improve their writing.	**0.55** **0.47** **0.50** *quantity* **0.40** *quality*	Graham and Perrin (2007) Graham et al. (2012) Goldberg, Russell and Cook (2003)
Writing about what they have read	Pupils are explicitly taught how to summarise in writing the texts they have recently read. A number of writing activities have been found to be effective in promoting writing skills (as well as improving reading comprehension): paraphrasing an original text to create a condensed summary (précis); analysing a text, to interpret the text's	**0.82** **0.40**	Graham and Perrin (2007) Graham and Hebert (2010)

Table 6.1: (*cont.*)

Summary	Overview	Effect sizes	Reference
	meaning, or describing their reaction to it in writing; writing notes (e.g., key words, ideas or phrases) which capture the essential information.		
Undertaking engaging pre-writing activities	Before beginning a writing task, pupils take part in structured tasks to plan or visualise the topic to be written about. Activities might include having pupils draw pictures relevant to the topic; write out a writing plan independently or in pairs or groups; read articles linked to the writing topic and discuss them before developing a writing plan, etc.	**0.54** **0.30**	Graham et al. (2012) Graham and Perrin (2007)
Studying models of writing	Pupils are given models of the kinds of writing that they will be asked to produce: e.g., argumentative or informational essays. Pupils closely study the structure of these models and attempt to incorporate the important elements of each model into their own writing.	**0.30**	Graham and Perrin (2007)
Teach and practice transcription skills	Skilled writers do not have to spend a lot of time thinking about handwriting, typing or spelling. Developing fluency with these skills frees up cognitive resources to focus on the writing process and effective communication of the key messages.	**0.55** **0.60** *fluency*	Graham et al. (2015)
Creating routines to ensure frequent writing	Pupils students must write frequently if they are to develop effectively as writers. This is about developing physical stamina as well as practicing cognitive aspects of the process and learning to compose different genres of writing.	**0.24** *quality*	Graham et al. (2015)

outcomes on writing tests when compared with the control classes, but the average impact is not consistent. Overall each approach is a reasonably 'good bet', and the higher the average effect size the greater the probability of success. However, the range also suggests that it is important to think carefully about which approach will make a difference in a new context. These are average differences; a teacher and a class may already be 'above average' in terms of an aspect practice or already doing better on these dimensions of writing than the typical setting included in the research studies. If this is the case, then further development or an increased focus, on, say, working collaboratively, may not bring about further improvement if the class are already used to activities like collaborative planning and peer review. It is important to recognise that one of the principal strengths of an experimental design, the control group or the 'counterfactual' condition which enables the comparison to be made, is also a potential weakness (as discussed in Chapter 4). For a successful intervention in a well-designed randomised controlled trial or experiment, we can be reasonably certain that the difference between the two groups results from the researched intervention. The design has strong causal inference. However, unless the experiences of the control group are clearly described, we do not really know what it is better than. This is often described as 'usual practice'. So, we know that the approach has worked (on average) compared with typical or 'average' practice. What teachers interested in applying findings really need to know is how like the control group their class is, as this could help them decide if the new approach is likely to be beneficial in their own setting.

Combining Reading and Writing Approaches

One further aspect of the evidence on reading and writing is worth considering. This is from a few meta-analyses which have looked at the relationship between writing and reading (Graham and Hebert, 2010), the benefits of balancing reading and writing instruction (Graham et al., 2017) and the value of writing for success in other subjects (e.g., writing to learn: Bangert-Drowns et al., 2004). Graham and Hebert's (2010) meta-analysis indicates that writing about material which has recently been read improves students' comprehension of this material and that teaching students how to write improves their reading comprehension (an effect size of 0.37 on standardised tests), as well as their reading fluency (0.66) and word reading. Other findings include that increasing how much students write enhances their reading comprehension (0.35). These findings provide empirical evidence consistent with theories about the value of writing to facilitate reading (e.g. Fitzgerald and Shanahan, 2000). Bangert-Drowns and colleagues' (2004) meta-analysis of 48 school-based writing-to-learn approaches indicates that that writing can have a small, positive impact on conventional measures of academic achievement with an overall effect size of 0.26. Two features of the studies were associated with greater effects; these were the use of metacognitive prompts and longer use of the approach. Two factors were also associated with smaller effects; these were the use of the strategy with pupils age 10–13 and the use of longer writing tasks. Graham and colleagues (2017) meta-analysis reviewed experimental interventions from pre-school to high school to investigate whether literacy programmes balancing reading and writing instruction strengthen both reading and writing outcomes. To be included in the review, no more than 60% of the teaching could be for either reading or writing alone. These approaches improved reading comprehension (0.39), decoding (0.53) and vocabulary (0.35). They also showed improvement for writing (0.37), particularly writing quantity (0.69) and quality (0.47), with some impact on the mechanics of writing (0.18). The findings indicate that literacy programmes balancing reading and writing instruction can be effective across reading and writing and that the two aspects of literacy can be learned together effectively.

In all of these meta-analyses there is a spread of effects, and all approaches were not equally effective for all outcomes. The particular details of their targeting for particular groups and ages of pupils combined with their use and development is clearly important. Overall, they do indicate that it is possible to teach reading and writing with emphasis on both outcomes successfully. This has implications for understanding the effectiveness and efficiency of different approaches. The separate approaches to reading or aspects of writing show what is effective and that improvement is usually possible and how gains can typically be achieved. What the separate approaches do not show is how efficient they are in achieving these outcomes. The indication from the three meta-analyses reviewed in this section suggest that this is an important area to pursue in further research. School time and the curriculum are always under pressure. We need better measures of efficiency and better indica-tions of precision in terms of who is likely to benefit so that we can help schools manage this balance.

Chapter Summary

Learning to read and write are at the heart of an education system; however, we should not forget an emphasis on spoken language and communication is also valuable. This is because, as we learn new skills and knowledge, we need both to understand and to articulate our learning. Speaking and listening activities also make aspects of knowledge and thinking public and explicit in classroom discourse. I think this has two advantages. First, the knowledge and thinking are modelled for other learners who can participate directly or vicariously by listening and observing. Second, the teacher has an opportunity to observe and evaluate what is being learned and discussed, so as to intervene and provide feedback, or adjust the focus of the learning. I also think that we should not forget the value of being taught to communicate effectively, that listening well and presenting ideas clearly and articulately are valuable educational goals in themselves.

The evidence is consistent that the teaching of reading should consider both decoding and comprehension. The evidence is clear that both approaches can be successful. However, this general picture does not provide a clear set of recommendations for what to do in the classroom for a particular child or class. I think we need better diagnostic information here to help match teaching approaches with the current capability of individuals and classes.

Overall the evidence on writing suggests that improvement is possible but that writing is a complex process which needs both explicit instruction and motivated practice. The complexity of skills and processes indicate that the emphasis in teaching needs to change systematically to ensure overall progress in the balance between transcription skills to ensure fluency and accuracy and the content of the writing according to genre and purpose.

The separate evidence across these three areas of literacy indicates each can contribute to the effective teaching and learning of literacy. There is a wealth of evidence about the effectiveness of different approaches. However, the evidence where approaches are combined is also important to bear in mind. This contributes to our understanding of efficiency across literacy teaching and learning and about which we still know relatively little.

Some of the inferential issues about meta-analysis are also important to acknowledge. Meta-analysis is not just about whether something 'works' or not. The evidence suggests nothing works in education for everyone equally. There is always a spread of effects. We need a good understanding of its strengths and weaknesses to interpret the conclusions. As we become more confident about a finding of average effects being secure, we should remember that we have also averaged the control group performance, so we may be less clear about what the approach is better *than*. We also need to bear in mind that meta-analysis usually tells us whether something is effective but not that it is efficient. There is extensive evidence in education suggesting that nearly everything can be improved. The challenge is to focus attention on what needs to be improved, without taking time and effort from other activities across the curriculum which might then suffer.

Final Thoughts and Reflections

The variation in effects is one of the main reasons that I think there is an important difference between evidence for policy and evidence for practice (see Chapter 8 for a further discussion of this). At the policy level, you need to know that the evidence shows that the effects are consistent and the impact is very high to mandate an approach for general benefit. It is only when you reach an effect size of about 1.4 that you might expect more than half of those involved to benefit substantially from the new approach. Otherwise you risk reducing the effectiveness of the best teachers as you improve the practice of those who are less effective. This creates a policy in which everyone regresses to the mean, and there is a risk that there is no overall improvement. However, evidence for practice can be more nuanced and targeted, specific to the school, the teacher and the pupils concerned. This argues for identifying teaching as a profession and having an expertise model of professional development and learning. Evidence for a particular educational objective or goal and information about a particular class are used to inform professional judgement about where improvement can be achieved. Expertise is not a popular idea in the world of the Internet or that of some contemporary politicians,[4] but the idea has important implications for understanding the growth of knowledge and skills over time (for both teachers and pupils) and the differences between 'knowing that' and 'knowing how'. There is a strong case for thinking about learning as developing expertise, whether for children or adults.[5]

Evidence from research can inform decisions about what to try to bring about improvement. It can give you a good bet about what has worked successfully elsewhere. It is usually only sought when there is a challenge to professional fluency. You don't look for an answer to something that is not a question. The problematic nature of the starting point, I think, is

---

[4] On 6 June 2016, in an interview with Faisal Islam on *Sky News*, Michael Gove's said: 'I think that the people of this country have had enough of experts ... '
[5] The August 2017 special issue of the *Journal of Philosophy of Education* (51.3) has a focus on 'Education and Expertise' edited by Mark Addis and Christopher Winch. The articles review contemporary understanding of expertise in relation to teaching and learning and particularly the important conceptual issues involved.

important. It indicates that the issue or area of practice can probably be improved, or at least professionally you think it ought to be improved. Here, evidence from research and what has worked for others in other contexts should provide a good starting point. The challenge of moving from propositional knowledge (knowing that), to making it work for you (knowing how) is not straightforward. It is my experience that it is the more effective teachers and schools who are usually interested in evidence-based practice. Ironically, they probably have the least to learn and may find identifying research-based solutions more difficult as such solutions need to work not just 'on average' but bring about improvement for those who are already above average!

7 Unpicking the Evidence about Parental Involvement and Engagement

Key questions
What is the evidence for the impact of parental involvement on educational outcomes for their children?
How can variation in effects build our understanding?

They fuck you up, your mum and dad.
They may not mean to, but they do.
They fill you with the faults they had
And add some extra, just for you.

Man hands on misery to man.
It deepens like a coastal shelf.
Get out as early as you can,
And don't have any kids yourself.

Philip Larkin *This Be the Verse* 1971

Introduction

This chapter contributes to an understanding of meta-synthesis by reviewing the evidence about ways that schools and parents develop working partnerships so as to improve outcomes for children. It draws on the evidence accumulated for the Toolkit about parental engagement and involvement with schools and has been developed from an earlier article in the *Journal of Children's Services* (Higgins and Katsipataki, 2015).

The wider context is the accumulating research evidence about the relationship between parental involvement and learning outcomes. By parental involvement I mean school, family and community partnerships in children's learning in school (Todd and Higgins, 1998). It is about understanding how this involvement can be supported through intervention with parents and schools and in particular the impact on children's engagement in school and their academic achievement. A general consensus emerged in the last quarter of the 20th century that such partnerships were not only desirable but had a positive impact on educational outcomes for children of these families. Since Lewis and

Vosburgh's (1988) meta-analysis of the effectiveness of kindergarten intervention programmes showed that parental involvement added significantly to the long-term impact of early intervention (with an effect size of 0.16), there has been a general consensus that parents play an important vital in promoting children's school success. There is rather less clarity, however, about how to identify specific practices that have the most influence on academic attainment and what the role of the school is in supporting the development of these practices and approaches with parents. School, family and community partnerships is a field that needs to be better understood to provide guidance for schools on how to build such partnerships to help improve learning outcomes for children.

Three recent reviews (Jeynes, 2012; Gorard and See, 2013; Van Voorhis et al., 2013) have all argued that parental involvement may indeed be beneficial for preschool and primary-age children, but these conclusions rest on evidence which is not conclusive due to the design and methodological quality of the studies in this field. In particular, the impact of *increased* parental involvement was not often rigorously tested. It is therefore difficult to make clear recommendations for practitioners. There is still much to learn about how best to engage with and involve parents so as to improve their children's educational achievement. As meta-analyses and systematic reviews have become more plentiful, there is a need for overarching reviews to aggregate findings across these reviews to address specific research questions. An 'umbrella' review tends to focus on a broad issue and highlights findings relevant to the central problem. This chapter may therefore be thought of as an exemplifying an 'umbrella' review, or a review of reviews, of the existing quantitative evidence about parental involvement in their children's education (Ioannidis, 2009). The key questions that this synthesis aims to answer are what is the quantitative evidence that parental involvement in school has positive cognitive or academic outcomes for children and how consistent and reliable is this evidence? Further implications for interpreting evidence of this kind from meta-analysis are also discussed.

Approach

As in the preceding chapters, the synthesis draws on the meta-analyses identified through systematic searching for the Sutton Trust-Education Endowment Foundation Toolkit, which has a focus on cognitive and academic outcomes for children achievable through intervention. It does not include, for example, reviews of parent training which focus only on behavioural or social change, or on outcomes for parents. It includes some studies of preschool involvement, particularly where there was follow-up impact for children of school age. The focus on interventions, rather than correlational studies, is because we need to know what schools and parents can do to improve outcomes, rather than just understand what behaviours by parents or by schools are associated with more successful outcomes, as these may not be directly causal. Overall the evidence summarised in the Toolkit suggests that parental involvement is beneficial, with an overall pooled effect of 0.22, but there is a wide range of effects (see Figure 7.1), so it is important to explore this variation further.

As described in Chapter 4, this approach can be viewed as an 'umbrella' review. This technique is more commonly used in the medical world to provide a rapid overview of evidence to inform practice. The technique arose out of the work of the Cochrane Collaboration (Grant and Booth, 2009). An 'umbrella' review tends to focus on a broad issue and highlights findings relevant to the central problem. The approach does have weaknesses. For example, the latest evidence may not be included because there is a lag between publication and inclusion in a review. The underpinning reviews may also have different aims and inclusion criteria, making synthesis challenging. However, it is particularly valuable to see patterns and gaps in the evidence and to provide an overview of the landscape.

Epstein (2009) identified six different types of parental involvement clustered around parenting, communicating, volunteering, learning at home, decision-making and community collaborations. Whilst these are hard to define precisely (see Jeynes, 2005) they are broad, helpful categories looking at both the activities and the relationships involved. For the analysis presented here, these have been reduced to three: general approaches to develop parent and school partnerships (which may include a number of components), specific family literacy interventions, and targeted interventions for families in particular need. The first two are typically school-led initiatives. The third category often has a broader focus than education and will typically come from a health or social care perspective, but it is included here because of the educational potential. Effect size gains are estimated as months of progress according to the Toolkit conversion with the rationale in the technical appendices, based on annual gains on standardised tests (Higgins et al., 2013). Effect sizes with confidence intervals and standard errors are also reported for comparison; the number of studies included in each meta-analysis is also provided.

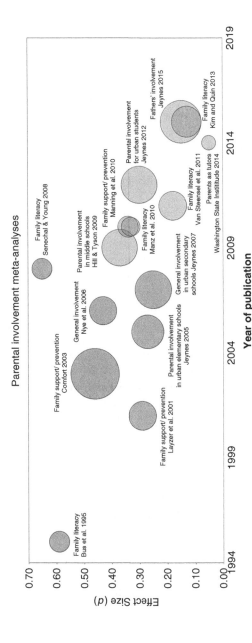

Figure 7.1: Bubble plot of parental involvement meta-analyses

General Parental Involvement Programmes

There are two main challenges in interpreting the reviews about general parental involvement programmes. First, most interventions have a number of components (such as parent workshops, meetings in school, volunteering opportunities or home activities). There are very few replications of evaluations of successful interventions, none which systematically vary the different components. This makes it difficult to identify the impact of each activity or of the different configurations of components. It also makes it difficult to compare the impact across programmes. Second, the design, implementation and analysis of evaluations vary and key aspects, such as attrition, are rarely reported (see Gorard and See, 2013 and Van Voorhis et al., 2013 for more discussion of this issue).

Overall the results from these five meta-analyses indicate that there is an important potential benefit from developing more effective partnerships between schools and parents, which ranges between three and six months' additional gain in academic outcomes for children receiving the intervention. A summary of the findings can be seen in Table 7.1.

It is useful to look down the fourth column of the table to see the analyses the researchers have conducted to see how the effects vary according to particular features of the different studies. These 'moderator' variables can be explored across the different studies to see if they help to explain the variation in effects. Two technical issues are important to note here. First, many meta-analyses do not have enough included studies with all of the information needed to draw clear conclusions. Second, this kind of analysis is based on an analysis of the patterns in the findings between studies. This kind of inference across meta-analyses is correlational, so we have to be very careful about drawing causal conclusions, even though the underlying studies may be randomised trials. One further point is worth making here. The majority of this evidence comes from North America, and Sénéchal and Young (2008) found studies from the US had a higher effect size, so some caution may be needed in generalising to other cultural contexts, particularly as there is often significant variation in impact related to ethnicity and socio-economic status. We know that these patterns are not necessarily consistent across countries and cultures (Kao and Thompson, 2003).

Overall the analysis shows that educationally important gains are achievable across the age range, with some indication of greater gains for older pupils (Nye et al., 2006; Jeynes, 2012). Impact can be seen across subjects but with more secure evidence for reading and literacy

Table 7.1: *General approaches to parental involvement*

Citation	Summary	Notes	ES of moderator variables
Hill and Tyson (2009)	**+ 5 months** **50 studies** **ES = 0.37** SE = 0.066 CI 0.24 to 0.49	Investigates parental involvement in middle school determine relationships with achievement. Five interventions in 50 included studies. Homework association negative. Overall r = 0.18 (CI 0.12 to 0.24); five interventions r = 0.19 weighted mean. Some associations dependent on correlational studies.	*Type of PI:* School based = 0.02; Home based = 0.03; Academic socialisation = 0.39; Help with homework = −0.11; Other activities = 0.12 African American = 0.11; European American = 0.19
Jeynes (2005)	**+ 4 months** **41 studies** **ES = 0.27** SE = 0.14 CI 0.00 to 0.54	Parental involvement and the academic achievement of urban elementary school children. Includes some correlational studies and analysis.	*Type of PI:* Parental expectations = 0.58. Parental reading = 0.42; Parental communication = 0.24; Checking homework = −0.08; Parental style = 0.31; Specific parental involvement = 0.29; Attendance of participation = 0.21; Sample: Mostly minority = 1.01; All minority = 0.41.
Jeynes (2007)	**+ 3 months** **52 studies** **ES = 0.25** SE 0.07 CI 0.11 to 0.39	Influence of parental involvement on educational outcomes of urban secondary school students. Measures: combined overall academic achievement, grades, standardised tests, and other measures (e.g., rating scales, academic attitudes and behaviours). Positive effects for	*Type of PI:* Parental expectations = 0.88; Parental style = 0.40; Parental communication = 0.24; Checking homework = 0.32; Specific parental involvement = 0.39 Rules = 0.02;

Table 7.1: (*cont.*)

Citation	Summary	Notes	ES of moderator variables
		both White and minority children. Includes some correlational studies.	Attendance and participation = 0.11 *Sample*: Mostly minority = 0.53 All minority = 0.42
Jeynes (2012)	**+4 months** **51 studies** **ES = 0.30** SE = 0.092 CI.12 to.48	Examines relationship between parental involvement programmes and academic achievement of pre-kindergarten to 12th grade. Includes studies with true control and some correlational.	*Age groups*: Younger = 0.29, Older = 0.35 *Type of tests*: Standard = 0.31, Non-standard = 0.21 *Specific Interventions*: Shared Reading = 0.51; Emphasised partnership = 0.35; Checking homework = 0.27; Communication teacher/ parent = 0.28; Head Start = 0.22; ESL training = 0.22
Nye, Schwartz and Turner (2006)	**+6 months** **19 studies** **ES = 0.45** SE 0.102 CI.25 to.66	Campbell review: parent involvement has a positive and significant effect on children's (5 to 10 years of age) overall academic performance with an effect large enough to have practical implications. Striking as the median length of parent involvement was only 11 weeks. Includes only randomised control studies.	*Academic outcome*: Reading= 0.41; Math: 0.54 (non-significant if outlier removed); Science = 0.08 PI and reward in math = 1.18; PI and parents training = 0.61 Longer duration not larger effects. b = 0.01.

than science or mathematics. There is contradictory evidence about for how long it is helpful to develop approaches, with longer interventions not necessarily showing greater effects (a median of 11 weeks appears optimal: Nye et al., 2006). By contrast, for literacy interventions, workshop programmes of more than five months duration were more effective, on average (about a month more progress (ES = 0.08): see Van Steensel et al., 2008). Frequency of contact or the intensity was hard to assess, but there was some evidence that shorter workshops (one to two hours) were more effective than longer sessions (three hours or more: see Van Voorhis et al., 2013). This level of complexity makes it difficult to give clear recommendations. Gains are possible, but implications of the findings about the targeting, the subject focus and the design of a programme of activities are much less conclusive.

There is considerable variation in the quality of the underlying studies. Overall, the field has relied on correlational and non-experimental designs that help us understand what successful parents do but not how to work in partnership to improve or develop the impact of what they *could* do. Studies rated of higher quality often had higher impact (0.04 higher: Jeynes, 2012). Nye and colleagues' review (2006) was conducted following Campbell Collaboration[1] quality procedures and identified overall a higher mean effect (0.45) with strict inclusion criteria. Some caution is indicated from this study's finding that journal publications reported a higher effect (0.63) when compared with non-peer-reviewed reports (sometimes known as 'grey' literature) as this may indicate publication bias, though other studies did not report significant variation by publication type (e.g., Jeynes, 2012). Overall, the meta-analyses investigating the impact of general parental involvement on children's learning shows a moderate effect, but as discussed earlier considerable caution is needed due to the variation in the quality of the underlying studies.

Home and Family Literacy Programmes

The five meta-analyses in this area found a range of average effects from two to eight months' additional progress in reading measures (see Table 7.2). The range is very broad and is likely to relate to the diversity of programmes, from book reading (Bus et al., 1995) to family literacy activities (Manz et al., 2010; Sénéchal and Young, 2008; Van Steensel et al., 2011) and to summer home reading programmes (Kim and Quinn, 2013), with many of these programmes targeted at low-income families. More information was available about longer-term impact, with more robust evidence of decline in

[1] https://www.campbellcollaboration.org

Table 7.2: *Family literacy programmes*

Citation	Summary	Notes	ES of moderator variables
Bus, Van Ijzendoorn and Pellegrini (1995)	**+ 7 months** **33 studies** N = 3,410 **ES = 0.59** *SE and CI not available*	Parent-preschooler joint book reading across several outcome measures. Explains about 8% of the variance in outcomes and affects acquisition of written language register. Effect not dependent on SES or on methodological differences. Effect smaller as children become readers and can read on their own.	*Type of PI:* book reading and language measures = 0.67; and emergent literacy = 0.57; and reading achievement = 0.55 Publication year: older studies have larger effects All others non-significant: sample size; publication status; SES: design; book reading measure; age at outcome measurement.
Kim and Quinn (2013)	**+ 3 months** **14 home interventions (41 studies)** **ES = 0.22** SE 0.031 CI −.03 to .48	Summer reading interventions conducted in US and Canada, 1998 to 2011. Classroom and home-based summer reading from K to grade 8, majority low-income children. Home interventions: mean reading 0.12; comprehension 0.22; fluency and decoding 0.26; vocabulary −0.02. Larger benefits for children from low-income backgrounds.	*Research-based:* yes = 0.25, no = 0.06; *Income:* majority low-income = 0.10; mixed-income = 0.08. *Design:* experimental = 0.09; non-experimental = 0.11. *Type of publication:* peer-reviewed journals = 0.11; unpublished = 0.13. *Timing of assessment:* immediate measures = 0.52; delayed measures = 0.20
Manz et al. (2010)	**+ 4 months** **14 studies** **ES = 0.33** SE 0.03 CI .27 to .39	Family-based emergent literacy interventions: Intervention studies involving caregivers for children (2–6 years) with an experimental or quasi-experimental design. Significant limitations in generalisability of this literature to these important groups of children.	*Sample characteristics:* Caucasian = 0.64; minority = 0.16; low-income = 0.14; middle/high income = 0.39 *Intervention type:* dialogic reading only = 0.32 *Intervention context:* home only = 0.47; home and school = 0.13; pre-intervention = 0.32; pre and during intervention = 0.33
Sénéchal and Young (2008)	**+ 8 months** **16 studies** N = 1,340 families **ES = 0.65** SE 0.061 CI .53 to .76	The effect of family literacy interventions on children's acquisition of reading from kindergarten to grade 3. Further analyses revealed that interventions in which parents tutored their children using specific literacy activities produced larger effects than those in	*Intervention characteristics:* parent involvement, read books to child = 0.18; listen to child read = 0.52; tutor child to read = 1.15. *Amount of parent training:* short (1–2 hrs) = 0.97, long (3–13.5 hrs) = 0.37. *Supportive feedback:* yes = 0.62, no = 0.70. *Length of intervention:* 1.5

Study		Summary	Characteristics
		which parents listened to their children read books. The three studies in which parents read to their children did not result in significant reading gains.	months or less) = 0.61; between 2.5 and 5 months = 0.57, 10 months + = 0.46. *Participant characteristics: Grade:* K = 0.51, grades 1–3 = 0.74. *Level:* normal = 0.69, special = 0.40. *SES:* low = 0.43, middle to high = 0.61. *Study characteristics:* *Design:* experimental = 0.67, quasi-experimental = 0.61. *Sample size:* <50 = 0.59, large >50 = 0.65. *Outcome measure:* early literacy = 0.46, word reading = 0.31, comprehension = 0.46, composite measure = 0.69. *Time of test:* immediate = 0.52, delayed = 0.79. *Country:* US = 0.78, non-US = 0.51. *Tests:* standard = 0.42, non-standardised = 1.24. *Publication year:* pre-1990 = 0.85, 1990 or later = 0.35
Van Steensel et al. (2011)	**+ 2 months** **30 studies** **ES = 0.18** SE 0.06 CI .06 to .30	Family literacy impact studies from 1990 to 2010; 47 samples, and distinguishes between effects in two domains: comprehension-related skills and code-related skills. A small but significant mean effect emerged (d = 0.18). There was only a minor difference between comprehension- and code-related effect measures (d = 0.22 vs. d = 0.17). No statistically significant effects of the programme, sample and study characteristics inferred from the reviewed publications. Children were between 2 to 10 years old.	*Activity type:* shared reading = 0.05, shared reading + other activities = 0.21, literacy exercises = 0.17. *Programme focus:* comprehension = 0.13, code = 0.16, both = 0.22. *Staff quality:* professionals = 0.21, semi-professionals = 0.18, both = 0.12. *Home visits:* yes = 0.18, no = 0.18. *Group meeting:* yes = 0.20, no = 0.12. *Book provision:* yes = 0.18, no = 0.18. *Location:* home-based = 0.17, home-based + centre-based = 0.24. *Duration:* < 5 months = 0.13, > 5 months = 0.21. *Educational status:* at-risk = 0.16, not at-risk = 0.20. *Age group:* pre-formal = 0.19, formal = 0.14, both = 0.26. *Sample selection:* random = 0.11, non-random = 0.22. *Pretesting:* yes = 0.15, no = 0.24. *Time of measurement:* short-term = 0.20, follow-up = 0.04.

follow-up measures (Kim and Quinn, 2013: 0.52 to 0.20) and washout (Van Steensel et al., 2011: 0.20 to 0.04) than increase (Sénéchal and Young, 2008: 0.52 to 0.79): see also Chapter 6).

Again, reviewing the moderator variables in the final column, there are few consistent messages. In terms of methodological features, one key issue is that older studies tend to have larger effects (Bus et al., 1995; Sénéchal and Young, 2008), perhaps reflecting the development of more rigorous approaches to evaluation in recent years or increasing provision and effectiveness of early years education. Other features such as design and publication bias appear less critical across these meta-analyses. There was contradictory evidence about duration, with shorter interventions sometimes having greater impact (Sénéchal and Young, 2008), sometimes longer (Van Steensel et al., 2011). On the other hand, shorter workshops with parents (an hour or so) are associated with larger effects (Sénéchal and Young, 2008).

Targeted Interventions for Families in Need

This final section looks at cognitive and academic outcomes for children in more specific need, through concerns about parenting or families in crisis. These targeted approaches can also support families with children who have special educational needs where more individualised support is indicated. Gains in cognitive outcomes of between four to six months, on average, are achievable. Indications are that frequency, intensity and duration of support are all important with reasonable consistency across the meta-analyses. Gains can be sustained, even into adolescence. Supporting younger teenage parents, especially with practical activities, is likely to be of benefit.

In terms of the quality of the studies in this category the most common observation is the lack of reporting of attrition in two out of the three studies (see Layzer et al., 2001; Manning et al., 2010). On the other hand, all studies presented information regarding the design of the studies, the sample characteristics and information about specific programme traits.

Across these meta-analyses, some of the moderator analysis is more consistent. Longer and more intensive interventions are linked with greater effects. Targeted interventions for those with greater need also appear more effective. Overall the indications suggest that for these children we should intervene early, intervene intensively and sustain the intervention over several years, ideally with a flow-through or follow-up component into schools. Although results are not guaranteed, this kind of targeted support to individual families can bring about significant short-

Table 7.3: *Parent and family support and intervention programmes*

Citation	Summary	Notes	ES of Moderator variables
Comfort (2003)	**+ 6 months** **94 studies** N = 6,147 **ES = 0.46** SE 0.041 CI 0.38 to 0.54 On cognitive/language outcomes Follow-up: d = 0.52 SE 0.041 CI 0.44 to 0.59	Effectiveness of parent training for children between the ages two and five to enhance child outcomes and examined variables related to the differential impact of parent training. When the theoretical orientation of programmes was considered, there was no evidence of differential effectiveness. Various instructional techniques used in parent training were not differentially effective, with the exception of some evidence of enhanced effect when a 'bug-in-the-ear' device was used.	*Design:* pre-post control = 0.66; random = 0.42; non-random = 0.57 *Type of sample:* universal = 0.17; selective = 0.17; indicated = 0.33; treatment = 0.50 *Sample source:* community = 0.25; referred/self-referred = 0.41 *Nature of problems:* externalising behaviour problem = 0.17; others = −0.09 *Orientation of training:* behavioural = −0.08; developmental = 0.47; other = 0.55. *Degree of intervention:* PT only = 0.49; PT and other = 0.37. *Format of training:* Individual families = 0.57; group = 0.52; individual and group = 0.23; self-instruction = −0.02. *Attrition:* 0–4% = 0.76; 5–24% = 0.31; 5% or greater = 0.30 *Total training time (in minutes):* 0–499 = 0.22; 500–999 = 0.31; 1,000 or greater = 0.53 *Role play:* no = 0.48, yes =.38 *Didactic:* no = 0.43, yes = 0.46 *Home visitation:* no = 0.51, yes = 0.46 *Modeling:* no = 0.51, yes = 0.42 *Video:* no = 0.43, yes = 0.43 *Homework:* no = 0.25, yes = 0.46
Layzer et al. (2001) (see also additional analysis in Sweet and	**+ 4 months** **260 programmes** **ES = 0.27** (across ages) d = 0.37 (preschool) *SE and CI not available*	Meta-analysis from final report of National Evaluation of Family Support Programs, findings from 260 programmes with representativeness compared with 167 family support programmes not evaluated.	*Randomised studies:* early childhood education: yes = 0.48, no = 0.25. Targeted to SEND children = 0.54, not targeted to SEND = 0.26. Peer support for parents = 0.40, no support = 0.25.

Table 7.3: (*cont.*)

Citation	Summary	Notes	ES of Moderator variables
Applebaum (2004)		All programmes providing family support services had small but statistically significant average short-term effects on child cognitive development and school performance, child social and emotional development, child health, child safety, parent attitudes and knowledge, parenting behaviour, family functioning, parental mental health and health risk behaviours, and economic well-being. Associated with stronger child outcomes were programmes that targeted special needs children. Home visiting as primary method of working associated with less strong child outcomes.	*Home visiting vs. parent groups:* yes = 0.26, no = 0.49. Home visiting SEND = 0.36, no SEND = 0.09. Parent groups SEND = 0.54, no SEND = 0.27. *Professional parent education staff vs. para-professional:* yes = 0.39, no = 0.23. *Case management provided:* yes = 0.08, no = 0.23. *Targeted to children developmentally at risk:* yes = 0.39, no = 0.22. *Serves majority low income families:* yes = 0.12, No = 0.22. *Teenage parents:* parent–child activities = 1.00, no parent–child activities = 0.50. *No teenage parent:* parent–child activities = 0.71, no parent–child activities = 0.21.
Manning, Homel and Smith (2010)	**+ 4 months** **17 studies** **ES = 0.34** SE 0.051 CI 0.24 to 0.44 Cognitive development	Meta-analytic review of early developmental prevention programmes (children aged 0–5: structured preschool programmes, centre-based developmental day care, home visitation, family support services and parental education) delivered to at-risk populations on non-health outcomes during adolescence (educational success, cognitive development, social–emotional development, deviance, social participation, involvement in criminal justice and family well-being).	Largest effect for educational success during adolescence (ES.53); followed by social deviance (0.48), social participation (0.37), cognitive development (0.34), involvement in criminal justice (0.24), family well-being (0.18), and social–emotional development (0.16). *Programme components:* 1 = 0.44, 2 = 0.44, 3+ = 0.42. *Programme intensity:* 500 min or fewer = 0.28, 500 or more = 0.49. *Duration:* 3+ years = 0.47, 1–3 years = 0.30. *Follow-through component:* yes = 0.51, no = 0.36.

term and sustained educational benefits for vulnerable children and young people. This can even be identifiable in adolescence with additional progress of seven months for prevention programmes in the early years. Support for families with children with special educational needs and disability appears particularly helpful, and for these families, home visiting appears particularly helpful.

Conclusions about Parental Involvement

The meta-analyses included in this 'umbrella' review allow us to draw similar conclusions to the more recent critical reviews of the field reviews (Jeynes, 2012; Gorard and See, 2013; Van Voorhis et al., 2013) which will be discussed later in this section. These reviews do provide a wealth of information but are somewhat different from the present review. More specifically, the first review (Jeynes, 2012) is a single meta-analysis of 51 studies focusing on the relationship of parental involvement programmes and academic achievement. The second, by Gorard and See in 2013, is a review including only primary studies investigating specific parental involvement programmes rather than overall impact, excluding meta-analyses. Finally, the Van Voorhis and colleagues review from 2013 includes a range of types of primary studies (descriptive, non-experimental and experimental) as well as meta-analyses. Some other differences between the reviews include specific inclusion criteria such as: focusing on older age ranges or using only published studies or including experimental evidence only. This review summarises the findings across these meta-analyses so as to consider the particular strengths and weaknesses of this overarching approach.

Three main conclusions can be drawn from the findings reviewed here. First, there is clear evidence of the potential of developing effective partnerships between schools and parents so as to increase children's educational attainment. Second, the variation in effects is complex and inconsistent making it difficult to make clear recommendations for policy and practice. Third, the design, analysis and reporting of studies makes it difficult to draw clearer conclusions and separate out the effects.

Some of the patterns of findings in moderator analyses are worth further exploration. For example, for schools, a programme of regular short but focussed workshops (an hour or so) over a limited period (10 weeks or so), which boosts parents' confidence and gives them practical activities they can undertake with their children in literacy or mathematics is likely to be a good starting point. It will be important to evaluate impact to be sure that the investment of time and effort is worthwhile, as these are correlational implications from this analysis rather than clear causal claims. There are some developing patterns in the findings, but there are also inconsistencies that we need to be able to understand and explain. The impact of homework, for example, appears to vary considerably. This may relate to the definition of homework, or it may be the range of ways that parents seek to support their children at home. This area has risks, as the effects of helping with homework (Hill and Tyson, 2009) and checking homework (Jeynes, 2005) are not always positive. Many teachers see sending reading books home as 'homework', but again there are some implications from this review. For preschool children who are not yet reading, supporting parents in reading *to* their children is important. But once children begin to read, the focus should shift to supporting parents in developing their children's reading capability and hearing them read. Different ages seem to require different approaches, if the goal is reading proficiency, so this is an important aspect to consider in future research.

Generalising or Offering 'Good Bets'?

We should also be cautious about generalising impact. For policy audiences, this means not promising too much, whilst drawing attention to the potential gains if successful interventions and approaches can be developed and implemented successfully. For practitioners, encouraging professionals to consider the evidence in making decisions is important. I think of these as 'best bets'. These are areas where other people have tried and, on average, have succeeded. However, practice in schools should not only focus effort on these high average impact areas. If schools are already engaged in activities which evidence suggests are less successful, on average (such as home visits), then these areas might benefit from a review. I'd call would call these 'risky' bets but would not necessarily suggest stopping them (unless you know you have something better to replace them with), but to review and ensure your use of the approach gives you an above average chance of success.

Chapter Summary

Overall, the evidence from these meta-analyses indicates that parental involvement, where school, family and community partnerships are developed to support and improve children's learning in school, offers a realistic and practical approach that has consistent evidence of beneficial impact on children and young people's attainment. Clear and specific messages for practice are hard to draw due to the nature of the evidence, its comparability and particularly the quality of the underlying studies.

A number of areas have promise can be identified by looking across the findings. Early literacy approaches are usually beneficial with as much as seven or eight months' additional progress achievable in terms of young children's learning, though we are less sure about how to consolidate these gains in the medium to long term. There are also other areas of practice, such as home visiting or parental support for homework, which, on average, are less successful. These are areas where practitioners may wish to review what they do to ensure the impact on learning is being achieved or to replace these approaches with others where the evidence indicates greater benefit is more likely. Further rigorous research and replication are also required, together with scale-up studies, to develop our understanding of the causal mechanisms for impact on learning outcomes. This is necessary to ensure any policy messages about parental involvement are likely to be successful. The analysis also highlights the importance of caution in drawing conclusions from meta-analysis which are not warranted by the strength of the underlying studies or which are based only on associations between studies where we are not sure of the causal pathways.

Final Thoughts and Reflections

My master's dissertation explored the use of mathematics games and activities provided by a school for use at homes in a disadvantaged community in North-East England. It was an ethnographic study, drawing on theories of discourse and interaction and the development of mathematical language and understanding. One of the things I learned was that the parents of the children I worked with all wanted their children to be happy and successful at school and that promoting mathematical activities, shared between home and school were possible and perceived as positive by all of those involved. The challenge was turning such activity into learning. It is not easy to translate shared experience into skills and knowledge that connect to the curriculum and help children's progress in mathematics. Supporting schools and families to improve children's mathematical skills and capabilities, whilst building mutual respect between home and school is difficult. Parental involvement should, of course, be valued above and beyond the instrumental gains for schools. As teachers, we educate children and young people on behalf of their parents and caregivers, and for the benefit of the young people themselves. It is all too easy to forget this and see involvement only as a means to improve school outcomes. I am not as pessimistic as the use of the quotation from Philip Larkin in the introduction to this chapter might suggest, but I do think one of the goals of education is to mitigate the 'misery man hands on to man'. In the UK, we have stronger intergenerational patterns of educational inequality and lower social mobility than in Australia or Canada. This is sometimes referred to as 'The Great Gatsby Curve' when you plot the relationship between inequality and intergenerational social immobility in several countries around the world (Corak, 2013). Other countries' cultures and educational systems suggest this is not inevitable, though changing these patterns is more complex than just improving education for children from disadvantaged backgrounds. Only part of the solution lies with schools and education. The gap, or Larkin's 'coastal shelf' quoted in the opening to this chapter, is too great for some children and too entrenched in some communities and cultures to be overcome by simple interventions in school, especially with current funding allocations. We only have the education system that, as a country, we are prepared to afford.

8   Conclusions and Final Reflections

Key questions
What conclusions can we draw about evidence for education from
    meta-analysis?
What are the limitations and benefits?
How might this evidence be improved?

Map-making had never been a precise art on the Discworld. People
tended to start off with good intentions and then get so carried away
with the spouting whales, monsters, waves and other twiddly bits of
cartographic furniture that they often forgot to put the boring mountains
and rivers in at all.

Terry Pratchett, *Moving Pictures*, 1990

Introduction

This chapter draws together the argument of the book as a whole about the
value of meta-analysis and meta-synthesis as a technique to understand the
evidence about effective teaching and learning. We have seen how the
research into the effects of different interventions and approaches forms
a broadly consistent and coherent picture of what makes a difference, on
average, on tested learning outcomes for pupils. You can improve outcomes for learners by teaching more effectively or more efficiently by
extending the time spent teaching and learning specific outcomes or by
increasing the intensity of the teaching and learning. Approaches like
providing feedback, when successful, improve efficiency by changing
what teachers and learners do to reach particular educational goals. One-
to-one and small group teaching increase the intensity of teaching. Other
areas like metacognition and self-regulation aim to share responsibility for
learners so that they can work out where to put effort or practice and help to
improve their own learning. This makes teaching more efficient. It is also
important to remember that these successful interventions and approaches
also use appropriate curriculum content and that, even in a general area like
metacognition, the subject content and detail is important. Skills and
knowledge in education are inextricably entwined.

The chapter also reviews some of the limitations in the evidence and the challenges in interpreting meta-analyses at this level. It considers issues related to effective teaching and learning and what we can reasonably infer from test results in an education system which has become dominated by assessment and testing. The chapter concludes with discussion about the challenge of applying findings in schools and the even greater difficulty of using research findings in policy, as well as what the next steps might be for synthesis of evidence in education.

The Importance of Aggregation and Comparison

Single studies are not sufficient in education. There is too much variation between contexts and settings and between schools, teachers and pupils, as well as the in application of educational concepts and ideas to be confident of the findings from a single study, no matter how robustly designed, implemented and analysed. A cumulative and comparative approach, pioneered by others and explored in the preceding chapters, is an essential tool in making progress in education research and to prevent the pendulum of policy changes swinging backwards and forwards each decade. Meta-analysis offers us the best way to get an overview of research findings of intervention studies in a specific area, such as phonics. It can also inform our understanding of literacy, by looking across the meta-analyses of phonics and reading comprehension or other areas of intervention research in reading. Understanding the relative value of different teaching and learning approaches, such as collaborative learning or the contribution of digital technologies, can help set findings from different fields in perspective. This synthesis of research provides a map of the field. The map might be relatively small scale, more like a page in an atlas or a Google maps view of a country, with the key geographical or educational features visible in the landscape. It may not provide us with a route map or a set of directions for a particular journey because these are dependent upon the precise starting point and destination we have in mind. However, this map can help orient us as we focus in on a particular educational goal.

Meta-synthesis as Creating a '*Mappa Mundi*'

To pursue this analogy, it seems likely to me that the current state of knowledge derived from meta-analysis in education is a bit like a medieval map of the world, a 'mappa mundi', where some areas are better known and more accurate, such as learning to read. These are the mountains and rivers mentioned in the quotation from Terry Pratchet's *Moving Pictures* at the beginning of this chapter. In other areas, the evidence is less secure, but still coherent and positive, such as about collaborative learning or small group teaching. There are also other areas of research, like the 'here be dragons' region of a *mappa mundi* (see Figure 8.1), there are mythical tales of learning styles, multiple intelligences and coloured lenses which cure all kinds of dyslexia[1] and which all appear to offer an educational panacea. These regions need further conceptual exploration and experimentation to identify what it is that might be effective within these ideas and practices then further testing to demonstrate any conclusive benefit. They may be completely mythical, or the myths may have some basis in fact. This metaphor captures my understanding of where we currently are in using meta-analysis to make inferences to understand teaching and learning. Much of the educational territory has been explore but not by surveyors and cartographers. We can learn from the narratives of the explorers, the travellers' tales of ethnography and biography, but I believe we can be more systematic than this and start to accumulate probabilities from these possibilities so as to increase the odds of successful educational decision-making. Hence my choice of the cover picture for this book. The distortions and inaccuracies produced by the aggregation of individual studies, with their varying designs, populations and measures, first up to the level of meta-analysis then again up to the level of meta-synthesis means that this picture is not yet as accurate or precise as we need to inform policy or practice as effectively or as efficiently as we would like.

It is, however, the only approach we currently have which will allow us to do this. The alternative is to say that inferences just are not possible across studies and across meta-analyses. If the variation between studies can all be explained by methodological features, then the particular outweighs the general. If this is the case or the findings are confounded by design and measurement artefacts, then we have to accept that we can only rely on tradition, judgement and experience in improving outcomes for children and young people. We can only know about the effectiveness of individual approaches but not their relative efficiency. Each generation of teachers will have to learn through apprenticeship and craft knowledge, but we will struggle to improve on the overall effectiveness of the educational systems we currently have in place. These systems will continue to be susceptible to the political pendulum and the whim or personal beliefs of ministers. I do not believe that this is inevitable, but the evidence base

---

[1] Galuschka et al., 2014

that we have and the analytical tools we can use today are still relatively crude, not least because they were not designed to create such a map.

More optimistically, if we can explain some of the variation between studies consistently with their pedagogical features, such as the length or intensity of the programme or specific activities or behaviours associated with greater effects which contribute to the overall impact then we can refine approaches and improve them based on earlier findings.

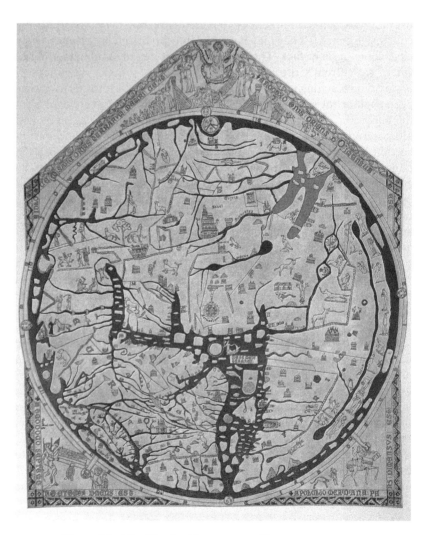

Figure 8.1: The Hereford 'mappa mundi'

Limitations in the Evidence

The meta-analyses used in the Toolkit and other similar approaches like *Visible Learning* (Hattie, 2008) and by other researchers such as Marzano (1998) were not conducted with meta-synthesis in mind. Meta-analysis is an average of averages, and the comparisons between these averages are then used to draw wider inferences. Meta-analysis combines the effects from different studies, which are already based on the average effects on groups of learners, to provide an overall average or 'pooled' effect. It also usually provides an indication of uncertainty, ironically called a 'confidence interval' around this average. We have to remember that this is an idealised or theoretical picture. What it is saying is that the overall pooled effect is the overall average, if you look across all of the studies included. So you have to decide how useful you think this is. How good is this average? How likely is it that this combination is helpful?

Although we have made progress in understanding what causes systematic variation (such as the age of pupils or outcome type which can be understood in relation to the measurement of effect sizes (as discussed in Chapters 2 and 4). Effect sizes from standardised tests are typically lower than researcher-designed measures, at least in part due to the measurement of treatment inherent measures (as outlined in Chapter 5). Other kinds of systematic variation have been identified, but the causes are less well understood. One example is the sample size, where larger studies report smaller effects (a correlation of -0.28, explaining about 8% of the variance: Slavin and Smith, 2009: p. 503). Whether this is publication bias or the impact of trial stage and type with pilot studies reporting higher effect sizes (Wigelsworth et al., 2016) or other aspects of 'super-realisation bias', is not yet well understood.

Certainly, aspects of research design can be linked with outcomes. However, randomisation, which controls for both known and unknown factors is not consistently associated with higher or lower effects. In the 1,000 or so meta-analyses in the Toolkit, for about a third randomisation is associated with higher effects, in another third with lower effects and in the remaining third it seems to make no difference. This needs further exploration, but my hunch is that when allocation bias or other potential issues with sample selection are hard to manipulate (or controlled for by the natural variation in school environments) randomisation may be a less important a factor to worry about, compared with other challenges such as attrition or measurement issues. Given the costs of large-scale randomised trials, there may be some circumstances where establishing impact can be done well enough with less rigorous designs, though I think randomisation will always be an important implement in the researcher's toolbox whether at micro- or macro-scale (Higgins, 2017), particularly to establish a robust causal pathway. Sometimes it is also important to find out that some things are not as effective as we believe so that we keep our aspirations achievable. Research can show us what not to do, or even what to stop doing, so as to focus on where improvement is more achievable. Education is a strongly optimistic endeavour, and as a primary school teacher I was driven by the triumph of the hope for my new class over the experience of the previous, year on year. Each new class was an opportunity to teach better than the year before in terms of helping my pupils to achieving their potential.

If we accept that meta-analysis is an important tool in accumulating research about educational interventions then we need to move forward more systematically than at present. Its potential has been acknowledged for more than 30 years in education (see Carol Fitz-Gibbon's articles in the *British Educational Research Journal* in 1984 and 1985 for example). For some questions, individual meta-analyses will of course be sufficient; others may require comparative inference. As indicated above, the next steps in ensuring greater accuracy are to explore methods to improve the rigour of these comparisons such as by developing network meta-analytic approaches (Cipriani et al., 2013) or creating a larger meta-analysis using the same inclusion criteria and common coding to ensure findings can be more directly compared.

The variation in impact across all of the areas of the Toolkit and found in other syntheses will never provide a precise prediction of what will be effective in any future application of research findings to a new context. Synthesis of this kind cannot therefore provide definitive claims as to 'what works' in education. Rather it is an attempt to provide the best possible estimate of what is likely to be beneficial based on existing evidence. Such studies summarise what has worked as an indication of, or perhaps a 'best bet' for, what might work in the future. Here the spread of findings should be taken into account, as well as the average. This indicates that findings from meta-synthesis should be used to inform practice and decisions about how to improve outcomes for learners, but the variation in findings in education and the lack of precision in the aggregation process means that applicability of this information to a new context is always going to be a probability rather than a certainty. It is always likely to need active enquiry and evaluation to ensure it succeeds in achieving the desired effects. This requires professional judgement and commitment to engaging with evidence on the part of the practitioner but also a disposition to interpret, challenge and test particular findings to ensure they are useful in a particular setting.

I think that the nature of education means that this will always be the case, so we need evidence-based probabilities to inform professional judgements about how to refocus teaching in relation to particular educational goals. An effective intervention with an effect size of 0.5 is only likely to provide additional benefit for 17% of the pupils in the intervention group, compared with the controls, who will still be making progress. This difference is what accounts for the gain for such an intervention. We need to know more about which pupils are likely to benefit from which interventions. At present, we know more about how to help learners who are falling behind to catch up with their peers than we do about how to accelerate whole classes or cohorts of pupils.

Testing and Educational Effectiveness

We rely on tests of learning to measure progress, and these kinds of assessment are an essential part of educational provision. They are also a vital aspect of research which seeks to identify impact. We need to measure that impact to estimate the benefit on an appropriate indicator, ideally which both captures the learning achieved and which will also provide a helpful indicator of future success. As a new primary school teacher, I used standardised tests of reading and mathematics and some Piagetian tasks in science to calibrate my judgements about the capabilities of individual children and the progress of my class. It was somewhat unfashionable at the time, but as an inexperienced teacher I wanted to know if my teaching was effective. I found that the baseline results at the beginning of the year provided a valuable range of diagnostic information for the range of capabilities in the class. I was not accountable for the results, and the head teacher was more curious about why I wanted to use such measures than in what they showed. My naïve assumption was that the children should make about a year's progress on such tests between early September and late July. In practice, though the overall progress was of this order (I tended to get better gains in mathematics than reading) what worried me was the spread of improvement and particularly the children whose progress appeared rather slower than I had hoped. I could often explain this in terms of the children's histories and experiences, but it was worrying that their educational trajectory did not seem to be building momentum. In reading, I had a clear image of a rocket on the launchpad, requiring intensive energy to get off the ground, then several booster rockets had to kick in at the right time to reach escape velocity, the transition when learning to read becomes reading to learn. Even if this speed is initially reached, for some children at the end of primary school their trajectory is still too low. Unless they receive additional help in the early years of secondary school they will be unable to read independently enough to support their own learning through texts across the subjects of the curriculum.

Over-reliance on testing is problematic in education for a number of reasons. I have outlined one dimension of this in what I think of as the 'proxy problem' in Chapter 6. My thinking here is influenced by the pioneering work of Donald T. Campbell. He was a psychologist and social scientist who used ideas from evolution in his understanding of social phenomena, and he had a particular interest in causation. He coined the term 'evolutionary epistemology' at both an individual and a cultural level. He also viewed all theorisation as provisional and fallible in the light of subsequent experimentation and testing[2]. He is also known for Campbell's Law: 'The more any quantitative social indicator is used for social decision-making, the more subject it will be to corruption pressures and the more apt it will be to distort and corrupt the social processes it is intended to monitor' (Campbell, 1976: p. 58). Campbell wrote further:

achievement tests may well be valuable indicators of general school achievement under conditions of normal teaching aimed at general competence. But when test scores become the goal of the teaching process, they both lose their value as indicators of educational status and distort the educational process in undesirable ways.' (p 61).

I think what depresses me most about this quotation is not its prescience, it's the date: 1976. This was more than 40 years ago. The corollary of this law is the impact of assessment and testing on the curriculum and on pedagogy, having three main effects. The first effect is the risk of a narrowing of breadth towards subjects which have high stakes tests, as the increase in time for preparation for English and mathematics performance squeezes out subjects like music, art and history. Cross-curricular themes and life-skills slip quickly between the cracks under such pressures, unless, like citizenship, panicked politicians of the day try to jemmy them back in to address a perceived crisis and artificially force them to survive in the curricular ecology, through the blunt instrument of

---

[2] This resonates strongly with philosophical approach known as pragmatism which I have an interest in. William James, the psychologist and physician, John Dewey the educationalist, and Charles Saunders Peirce, the logician, are recognised as its founders. For Peirce, science is not systematic knowledge, but a 'life devoted to the pursuit of truth according to the best-known methods on the part of a group of men [sic] who understand one another's ideas and works as no outsider can. It is not what they have already found out which makes their business a science; it is that they are pursuing a branch of truth according, I will not say, to the best methods, but according to the best methods that are known at the time' (Peirce, 1905). Peirce was one of the first to articulate the benefits of blinded, controlled randomised experiments in 1884 (Peirce and Jastrow, 1885).

inspection. We rarely consider the opportunity cost of marginal gains in tested achievement and the lost curriculum opportunities as a consequence.

The second effect is the narrowing within subjects towards the assessment points. The metaphor I have in mind here is it is like measuring the depth of a swimming pool using a stick to determine its depth. If you know the length and breadth, you can calculate the volume. However, if you know the measurement points, you can dig little holes underneath where the stick goes and make it appear that the pool has become deeper. This is the outcome of the 'proxy problem' in relation to Campbell's Law and specific test items. The measurement remains reliable, but the validity of the assessment has been compromised. What makes this more complex is that some of the functions of the assessment process in terms of it providing a ranking or sorting system are still functional, so the consequences may not immediately become apparent, particularly as the changes affect cohorts of students year by year and tests are normed for these cohorts. This affects educational research and meta-analysis in particular as one of the impacts of within-subject and curriculum stenosis is to inflate performance on the tests which are closely aligned with high stakes measures, particularly in subjects like reading and mathematics. Evaluation researchers depend upon these as measures of learning, but which are also good predictors of future educational achievement. I'd rather testing was used more for research and diagnostic purposes and less for accountability and summative judgements. The tests would retain their validity better, and we could use research to check that policy-based improvement was genuine.[3]

The final effect is on pedagogy where the emphasis shifts towards demonstrating performance, rather than broader knowledge, understanding and capability. An example here is the way pupils are prepared and practised to answer examination questions so that all of the marks they can score on the assessment rubric are overtly ticked in their answer but sometimes at the cost of the meaning and content, and certainly in terms of creativity and breadth of knowledge. If it came to a choice, I'd rather emphasise validity in assessments for practice in schools and risk reducing their reliability. This would help drive practice in schools towards learning, rather than performance. I'd reserve measures with greater reliability for research (which would include representative national sampling of standards of performance). The precision required

---

[3] Peter Tymms's 2004 article 'Are Standards Rising in English Primary Schools?' compared statutory test data in England with the results from several different research studies. He concluded that up to the year 2000 there were clear rises in standards which corresponded with, but were not as strong as the official data suggested.

for reliability comes with costs, one of which is that the validity is likely to be subverted, over time, either at the system level through curriculum and subject stenosis or by the strategic gaming of the assessment system and greater efficiency at test preparation and performance. Just to be clear, and to avoid being misquoted or mis-tweeted, of course I want an education system which promotes a high level of fluency in reading, writing and mathematical capability but not at the cost of depth of learning in these high stakes subjects and at the cost of the evaporation other subjects from the timetable. Education is not a time trial where shaving off tenths of a second for a cycling team with a new helmet are worth a significant investment of time and money. This kind of improvement comes at too high a curriculum cost for pupils. If the cyclists and their coaches had to design, make and test their own helmets and racing gear they would not be able to spend as much time actually cycling and training to ride faster. In the same way in schools, the teaching of precision semicolon placement displaces other teaching. Is it really worth this cost?

Why What's Worked Will Work (Probably)

In an article in 2007, Gert Biesta set out a clear critique of the ways in which evidence-based practice in education it has been promoted and implemented using the catchy title 'Why "What Works" Won't Work'. He draws attention to a number of important issues, in particular the dynamics between scientific and democratic control over both educational practice and research. He identified the 'scientific' with a 'technological model of professional action'. I argue that this is not a necessary connection, however, and the dichotomy is a false one. It is possible to hold a 'scientific' view of causation and at the same time see education as a process of symbolically mediated interaction. You can hold scientific beliefs and value the importance of people as individuals. For me, this means putting greater emphasis on the internal validity of findings from trials and aggregating the effects through meta-analysis so as to answer the question of how good a 'bet' a particular approach is. The subsequent question of 'And will it be effective in my school, for my pupils?', in terms of external validity, requires, in my view, either extensive replication to understand the range of contexts where it can be successful or professional judgement and interpretation based on the particular inferences meta-analysis can provide. We need to understand clearly what the causal processes are which need to be enacted in the new setting. This perspective is supported by a more rigorous understanding of what an 'average treatment effect' means in scientific terms, as discussed in Chapters 4 and 5 (and see, for example, Deaton and

Cartwright, 2016). This has clear implications for what can be inferred from the findings of randomised trials or other experimental research. This is in terms of what the average impact means in relation to the schools, teachers and pupils in the research and then how comparable these effects are in the studies included in a meta-analysis and then how similar these pupils, teachers and schools are to the context for 'evidence-use'. A stochastic or probabilistic understanding of causation is needed, rather than a deterministic one, but this has always seemed to me to be a requirement, given human complexity and agency.

Biesta (2007) examines three key assumptions of evidence-based education: first, the extent to which educational practice can be compared to the practice of medicine; second, the role of knowledge in professional action, particularly in terms of what kind of knowledge assumptions are appropriate for and relevant to professional practices that can be informed by research outcomes; and third, the expectations about the practical role of research implicit in the idea of evidence-based education. Perhaps unsurprisingly, I disagree with Biesta in some important respects on each of these issues, but most importantly his view that scientific knowledge diminishes democratic control over education and the decision-making of practitioners. By contrast, I argue that access to and engagement with 'scientific' knowledge is an essential condition for the democratic participation of teachers in making judgements about educational practice. I think it is naïve to think that education is not in competition with other areas of public life and the logic of scientific enquiry will trump other valid but less reliable and predictive ways of knowing in relation to causal arguments. For me it is therefore important that this evidence base is accessible to teachers and other education professionals so that 'scientific' research can be marshalled along with professional judgement to defend professional practice.

Biesta draws the conclusion that the notion of evidence-based practice is a limiting concept which not only restricts the scope of decision-making to questions about effectiveness but also that it restricts the opportunities for participation in educational decision-making. He argues that we must expand our views about the interrelations among research, policy and practice to keep in view education as a thoroughly moral and political practice that requires continuous democratic contestation and deliberation. On this point we agree, though I go further and argue that the role of evidence is an essential ingredient in this debate. Unless we understand likelihood of outcomes from educational practice in terms of how predictable they are, this debate will be limited to opinion and belief about educational goals which involve intention and causation. Evidence will always be necessary, but it will never be sufficient.

The implications about the uncertainty we currently have about the applicability of the current evidence base indicates to me it is always going to be more informative for practice than it can ever be for policy. This means that the role of evidence-based policy should be to support evidence-based (or more precisely evidence-informed) practice. This is because of the variation in findings across educational trials and the challenge of interpreting average treatment effects from single trials, as well as the extent of the pooled averages from research syntheses such as those found in meta-analysis. A successful educational intervention probably only benefits only a relatively small proportion of those involved, typically 10% to 30% of pupils (effect sizes between 0.3 and 0.8). However, as I argued in Chapter 4, in meta-analysis we have also averaged the counterfactual: we have averaged the performance of the control groups. So, the inference is that this has worked, on average, *compared with* the average. My further inference is that the better a school is performing the less likely it is that 'average' gains will be enough, as argued in Chapter 6. Their rate of progress is likely to be better than the average control group. We need to remember that the control group are making progress too. Here you'd need to identify more precisely what could help, by understanding the current capabilities of learners (what I think of as diagnosis[4]) or go for approaches with higher average effects which will offer greater potential for improvement. The issue for policy is that we need to remember that we are thinking of educational growth and progress over time and that a policy which benefits those performing less well, applied universally, may slow the progress of those doing better, holding them back from the progress they would otherwise have made. It is not like prescribing statins to reduce deaths from heart attack where the population benefit is small (an effect size of 0.018 (Baigent et al., 2010) benefiting about 1 in every 200 people taking statins), but the cost is very low. The 199 people taking statins who do not benefit[5] (on average) in this way are not prevented from making progress towards a health and well-being goal, in the same way that an educational intervention may alter the progress of those involved. The financial and

---

[4] Diagnosis is a word which can alienate people as it conjures a view of illness and deficit from the medical world. This can be helpful when thinking of remediation, but in general what we do in education is more analogous to promoting health and well-being. The diagnosis is of current capacities so as to develop educational capability further. I think one of the primary responsibilities of teachers is to undertake such a diagnosis for their class and the individuals within it, in relation to what the pupils can do and the educational aims or goals that the teacher (and society) has.

[5] There are other benefits from statins, such as reduced frequency of harmful cardiovascular events, reduced incidence of strokes and better recovery after non-fatal cardiac arrest. Statins also have some side effects such as muscle pain, headaches and nose bleeds.

opportunity cost of taking statins is minimal. There is always an opportunity cost of any educational intervention, as there is for all teaching choices. A child spends about 1,000 hours a year in school, so we need to think about any innovation as replacement. It is as important to decide what to stop doing as well as what to start. What else could the teacher and pupils be doing that would be more efficient or more effective in achieving educational goals?

A Model for Effective Research Communication and Use

Evidence alone is also not enough to change practice. Dagenais and colleagues (2012) conducted a systematic and comprehensive review of what seem to be the determinants of the use of research-based information by school practitioners. They classified research use as *general* or *local* and identified a number of features from the studies they reviewed in terms of the characteristics of the research, the communication processes, the practitioners and the schools involved. The team argue that these dimensions are central to understanding the factors that affect practitioners' decisions to become involved in change processes connected with evidence use and that also help to sustain their involvement in order to develop professional performance (see Table 8.1).

In 2014, the EEF committed £1.5 million to five projects to explore the impact of research-use in schools, evaluated through impact on pupils' attainment. Early indications suggest that although these approaches to promote research use may have promise, particularly in terms of developing awareness and engagement, it is challenging to demonstrate a causal link between evidence use and improved outcomes (Speight et al., 2016). As Langer, Tripney and Gough (2016) noted in their review, we still lack evidence on the relative effectiveness different approaches. There is some evidence that simpler and more defined interventions have an increased likelihood of immediate impact (e.g., Levin, 2011), so in the short-term research, like the EEF research-use projects, which improve our understanding of simple causal mechanisms may help to design larger studies of more complex interventions 'whose causal chain is difficult to disentangle at this early stage of research knowledge' (Langer et al., 2016, p. 4). In the context of the Toolkit, I talk about a model of use which helps to outline the conditions needed for effective communication of research findings which can benefit practice. The dimensions, however, are in tension with each other and improving one may compromise another.

Table 8.1: *Dagenais's research characteristics*

Research characteristics	Characteristics of communication	Practitioners' characteristics	School characteristics
Accessible and timely	Facilities	Skills and competencies	Enjoys external support
Objective and true	Access to research and data	Prior participation in research	Wants evidence for decision-making
Easy to understand and implement	Quality	Attitudes towards research	Encourages and supports initiative
Connected to school/ classroom context	Collegial discussions	Willingness to innovate	Has prior experience with initiatives
Relevant	Collaboration with researchers	Self-efficacy and commitment	Staff capacity and support to use research
	Sustained collaboration via networks and partnerships	Experience	Encourages internal collaboration
	Media	Prior coursework in research methods	Prioritises appropriate professional development activities
		Content area taught	Needs innovation
		Training on how to make use of research	Is committed to organisational learning
		Involvement in research	Allocates time and resources, including available technology

(adapted from Dagenais et al., 2014: pp. 297–299)

Some of the responsibilities in the model are from the perspective of the researcher. These involve the research being *accessible, accurate* and *actionable*. This immediately sets up a series of tensions for the researcher in summarising findings accurately but succinctly in a way which educational practitioners can understand and put into practice. Accuracy refers mainly to how findings are summarised in relation to what was found (the internal validity answering the question 'did it work there?'). There are also implications for external validity, as I mentioned in Chapter 4. I think this is problematic in education as we have almost no replication and the samples of schools, teachers and pupils are not randomly selected. It is

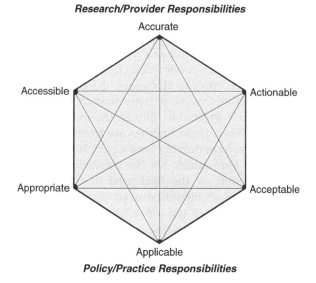

Figure 8.2: A model of research communication and impact

therefore hard to estimate the probability of an approach being successful somewhere new. We know that accessibility is a key issue both in terms of getting hold of research evidence but also in terms of understanding it. One of the main drivers behind the development of the Toolkit website was to create an *accessible* but *accurate* resource for education professionals (see Figure 8.2). Academic journal articles are constrained by genre, form and the history of the discipline and are rarely an easy read for the busy teacher. How you distil findings into *actionable* steps is even more challenging. I've likened this before to picking the strawberries out of the jam (Higgins and Hall, 2004). You can sometimes see that the fruit is there, but they are so boiled and crystallised by the meta-analytic jam-making process that they no longer taste like strawberries. Another challenge is that in a meta-analysis you may see patterns of effects suggesting the length of time that is most promising or the kinds of training which appear beneficial, but it is important to remember that even when these are based on rigorous randomised trials they are still associations. There may be a good causal claim from individual studies, but we do not know that we have identified the correct causal pathway by looking at the similarities across studies.

From the practice perspective, there are also responsibilities in terms of the research being *applicable, appropriate* and *acceptable*. A good fit between research evidence and the practice context is essential, and this needs to be the responsibility of the teacher or school to meet the educational needs and capabilities of their pupils. It is important to know that it is likely to be *applicable* in terms of subject, age and approach. One of my worries about research from other fields is being clear about how it might apply in classroom and curriculum contexts. Psychology research on motivation or neuroscience research into brain function are not directly applicable, though the findings often appear seductively suitable. When these findings are tested in the classroom they often do not always have the effects expected. It is also important to identify whether it is appropriate for the particular teacher and the pupils involved. To increase the likelihood of it being *appropriate* I think it needs to meet an identified need or a perceived problem, rather than being picked from the top a list of effective strategies or plucked at random from successful research findings. Identifying a problem or challenge is more likely to create a match between the research context where it successfully made a difference and the new setting. One way of looking at the Toolkit is that it is a compendium of solutions to educational challenges and the distribution of effects gives you a probability of how likely it is to be useful. The problem is that the questions to which we have these solutions are no longer attached. It is therefore important to consider whether a particular research-based practice is *appropriate* as a solution to the challenges a particular school or teacher faces. The final practice dimension is how *acceptable* the findings are. At one level, they have to be educationally acceptable. Some kinds of behavioural change may be very efficiently achieved with pain or discomfort, but they would not be educationally or ethically acceptable. The next level is more difficult to tackle, in that to stand a chance of being successful research findings have to be *acceptable* to the teachers involved. If the findings conflict with deeply held beliefs about effective practice, then they may either be rejected and not tried or adopted resentfully and set up to fail. I've always argued that, as a classroom teacher, if you presented me with a robust and rigorously researched reading intervention which was consistently successful when evaluated, that I could guarantee to make it fail in my classroom. Teachers are the gatekeepers of their own practice. The irony I've experienced here is that the teachers who are more open to research-based approaches are often the ones who are already highly effective. They actively seek to increase their repertoire of strategies and are keen to try out approaches backed by research. By contrast, sometimes those I've felt might benefit from trying out such strategies are the ones most likely to find them unacceptable. There are many reasons why the pharmaceutical model of research does not apply in education, not least of which is that it is not clear who is supposed to take the medication. The 'tablets' have to be palatable to the teacher yet be effective with the pupils.

Looking to the Future

Overall the Toolkit aims to provide *accessible* and *applicable* information by summarising research evidence from meta-analysis which is as *accurate* as current methods allow, but also sufficiently *actionable* to be useful in the classroom. There are certainly limitations to this approach, particularly in the balancing of the accessibility of the information, with its accuracy in drawing on research evidence, but also in providing sufficient detail and specificity to support action in the classroom. The key point is that this kind of approach the best option that we currently have to provide comparative findings that let us think about the relative benefits and costs. If we don't do this deliberately, we are accepting the existing cost/benefit conditions, implying we already think teaching and learning in schools are efficient enough. The alternative is to ignore or discard such findings from previous research as incommensurable. My own view is that we have to start somewhere and build the evidence-base into a more coherent map. The EEF is using the Toolkit, as well as other research, as a guide to commission further research, over 140 randomised trials between 2011 and 2017 involving over nearly a million pupils in schools and colleges in England. The evidence will both test the existing meta-analytic findings and generate further findings which will feed back into the evidence-base to increase its robustness. This approach will also create a database of findings from EEF trials where there is greater comparability across context and outcomes. This should help to identify more precisely the causes of variation in research findings or at least indicate what progress is possible in this area. Ideally, we would link in the developmental literature, subject research, international surveys and wider studies of perceptions and experiences to put some flesh on the bones of the experimental data, but it might take some time to turn the map into a digital global model of education research!

The inferences we have from meta-analysis are the best we currently have and will remain so until we increase the frequency of replication studies and collect data on the medium- and long-term impact of interventions, as well as the impact across a wider range of educational outcomes. In the meantime, we can start to create larger meta-analyses, with common inclusion criteria and coding, either by taking the ones we have and unzipping then and recombining into a larger overall meta-analysis or by systematically extending a database of single studies. This would not only help methodologically in understanding design and measurement effects better but would enable bespoke meta-analyses by subject and age which would increase the applicability of matching findings to context.

As a new academic, just out of the classroom, I was frustrated when I found convincing research evidence which contradicted practices I had been encouraged to adopt in the classroom, such as the whole language and 'real books' approaches that I mentioned in Chapter 1. Other research I found into the effective teaching of mathematics I had no idea existed but could have made me a better teacher in the classroom. I have been determined to find ways to make research more accessible to practitioners. I have tried not to be side-tracked from this educational evidence map-making, though it has certainly never been a precise art, either on the Discworld or in education, as the introductory quotation to this chapter indicates. The 'spouting whales, monsters, waves and other twiddly bits' of educational 'furniture' are indeed enticing, but I think the meta-analysts and meta-synthesists have started to put in the 'boring mountains and rivers', though there is still a lot of blank space on the map.

Appendix A   Details of the Meta-analyses Included

Appendix A.1.1: *Phonics meta-analyses*

Citation	Overall ES	Abstract	Moderator variables
Camilli et al. (2003)	0.24 SE 0.07	Examined the findings of the 'Teaching Children to Read' study of the National Reading Panel and the procedures of the study. Meta-analytic techniques found that the methodology and procedures were not adequate. Findings suggest that phonics, as an aspect of the complex reading process, should not be overemphasised.	Standardised tests yield larger effects. Studies were control groups used language approaches had lower effects and treatment using language approaches had larger effects.
Ehri et al. (2001)	0.41 SE 0.03	A quantitative meta-analysis evaluating the effects of systematic phonics instruction compared to unsystematic or no-phonics instruction on learning to read was conducted using 66 treatment–control comparisons derived from 38 experiments. The overall effect of phonics instruction on reading was moderate, d = 0.41. Effects persisted after instruction ended. Effects were larger when phonics instruction began early (d = 0.55) than after first grade (d = 0.27). Phonics benefited decoding, word reading, text comprehension and spelling in many readers. Phonics helped low and middle SES readers, younger students at risk for reading disability (RD), and older students with RD, but it did not help low achieving readers that included students with cognitive limitations. Synthetic phonics and larger-unit systematic phonics programmes	**Time of post-test: end of training** = 0.41, end of training or first year = 0.44, follow-up = 0.27. **Outcome measures:** decoding regular words = 0.67, decoding pseudowords = 0.60, reading misc. words = 0.40, spelling words = 0.35, reading text orally = 0.25, comprehending text = 0.27. **Grade level:** kindergarten and first grade = 0.55, kindergarten = 0.56, first grade = 0.54, 2nd–6th grade = 0.27. **Grade and reading ability:** kindergarten at-risk = 0.58, first normal achieving = 0.48, first at-risk = 0.74, 2˜6 normal = 0.27, 2nd–6th low = 0.15, reading disabled = 0.32. **SES:** low = 0.66, middle = 0.44, varied = 0.37, not given = 0.43. **Type of phonics programme:** synthetic = 0.45, large phon. units = 0.34, miscellaneous = 0.27, direct instruction = 0.48.

produced a similar advantage in reading. Delivering instruction to small groups and classes was not less effective than tutoring. Systematic phonics instruction helped children learn to read better than all forms of control group instruction, including whole language. In sum, systematic phonics instruction proved effective and should be implemented as part of literacy programmes to teach beginning reading as well as to prevent and remediate reading difficulties.

Instructional Delivery Unit: Tutor = 0.47, small group = 0.43, class = 0.39.
Type of control group: basal = 0.46, regular curriculum = 0.41, whole language = 0.31, whole word = 0.51, miscellaneous = 0.46.
Assignment of participants to groups: random = 0.45, non-equivalent = 0.43.
Existence of pre-treatment group differences: present = 0.13, absent = 0.47, present but adjusted = 0.48, not given = 0.40.
Sample size: 20–31 = 0.48, 32–52 = 0.31, 53–79 = 0.36, 80–320 = 0.49.

| Galuschka et al. (2014) | reading g = 0.32 [0.17, 0.46] spelling g = 0.33 [0.06, 0.61] reading adjusted for publication bias = 0.19 [0.03, 0.35] | Children and adolescents with reading disabilities experience a significant impairment in the acquisition of reading and spelling skills. Given the emotional and academic consequences for children with persistent reading disorders, evidence-based interventions are critically needed. The present meta-analysis extracts the results of all available randomised controlled trials. The aims were to determine the effectiveness of different treatment approaches and the impact of various factors on the efficacy of interventions. The literature search for published randomised controlled trials comprised an electronic search in the databases ERIC, PsycINFO, PubMed and Cochrane, and an examination of bibliographical references. To check for unpublished trials, we |

For reading outcome
reading severity:
mild = 0.44
moderate = 0.22
severe = 0.30
Amount:
up to 14 hours = 0.35
between 15 and 34 = 0.11
35 hours and more = 0.37
Duration:
up to 12 weeks = 0.26
more than 12 weeks = 0.35
Setting:
computer with teacher = 0.36
single subject = 0.20
Group = 0.37
Conductor:
study author = 0.8

Appendix A.1.1: (*cont.*)

Citation	Overall ES	Abstract	Moderator variables
		searched the websites clinicaltrials.com and ProQuest and contacted experts in the field. Twenty-two randomised controlled trials with a total of 49 comparisons of experimental and control groups could be included. The comparisons evaluated five reading fluency trainings, three phonemic awareness instructions, three reading comprehension trainings, 29 phonics instructions, three auditory trainings, two medical treatments and four interventions with coloured overlays or lenses. One trial evaluated the effectiveness of sunflower therapy, and another investigated the effectiveness of motor exercises. The results revealed that phonics instruction is not only the most frequently investigated treatment approach but also the only approach whose efficacy on reading and spelling performance in children and adolescents with reading disabilities is statistically confirmed. The mean effect sizes of the remaining treatment approaches did not reach statistical significance. The present meta-analysis demonstrates that severe reading and spelling difficulties can be ameliorated with appropriate treatment. In order to be better able to provide evidence-based interventions to children and adolescent with reading disabilities, research	student = 0.4 teacher = 0.24 special education specialist = 0.25 **Spelling/writing:** included = 0.15 not included = 0.33 **For spelling outcomes** **Reading severity:** mild = 0.41 moderate = 0.15 **Amount:** up to 14 hours = 0.43 b 15 and 34 = 1.14 35 hours and more = 0.05 **Duration:** up to 12 weeks = 0.17 more than 12 weeks = 0.31 **Setting:** single subject = 0.48 group = 0.26 **Conductor:** student = 0.94 teacher = 0.09 special education specialist = 0.14 **Spelling/writing:** included = 0.37 not included = 0.33

		should intensify the application of blinded randomised controlled trials.	
Jeynes (2008)	0.30 SE 0.10	This meta-analysis of 22 studies examines the relationship between phonics and the academic achievement of urban minority elementary school children. Further analyses distinguish between those studies that are of higher quality than the others and those studies that examine all minority students and mostly minority students. Results indicate a significant relationship between phonics instruction and higher academic achievement. Phonics instruction, as a whole, is associated with academic variables by about 0.23 to 0.33 of a standard deviation unit. This relationship holds for studies that examine all minority students and those that include mostly minority students. The results also hold for higher quality studies. The significance of these results is discussed.	**Age:** younger elementary = 0.25, older elementary = 0.30.
McArthur et al. (2012)	word reading accuracy = 0.47 non-word reading accuracy = 0.76 word reading fluency = −0.51 reading comprehension = 0.14 spelling = 0.36 phonological output = 0.38	Around 5% of English speakers have a significant problem with learning to read words. Poor word readers are often trained to use letter-sound rules to improve their reading skills. This training is commonly called phonics. Well over 100 studies have administered some form of phonics training to poor word readers. However, there are surprisingly few systematic reviews or meta-analyses of these studies. The most well-known review was done by the National Reading Panel (Ehri, 2001) 12 years ago and needs updating. The most recent review (Suggate, 2010) focused	The amount of evidence for each outcome varied considerably, ranging from 10 studies for word reading accuracy to one study for non-word reading fluency. The effect sizes for the outcomes were: word reading accuracy standardised mean difference (SMD) 0.47 (95% confidence interval (CI) 0.06 to 0.88; 10 studies), non-word reading accuracy SMD 0.76 (95% CI 0.25 to 1.27; eight studies), word reading fluency SMD −0.51 (95% CI −1.14 to 0.13; two studies), reading comprehension SMD 0.14 (95% CI

Appendix A.1.1: (*cont.*)

Citation	Overall ES	Abstract	Moderator variables
	letter-sound knowledge = 0.35 non-word reading fluency = 0.38	solely on children and did not include unpublished studies. Objectives The primary aim of this review was to measure the effect that phonics training has on the literacy skills of English-speaking children, adolescents and adults whose reading was at least one standard deviation (SD), one year or one grade below the expected level, despite no reported problems that could explain their impaired ability to learn to read. A secondary objective was to explore the impact of various factors, such as length of training or training group size, which might moderate the effect of phonics training on poor word reading skills. Search methods and selection criteria We included studies that use randomisation, quasi-randomisation or minimisation to allocate participants to either a phonics intervention group (phonics alone, phonics and phoneme awareness training, or phonics and irregular word reading training) or a control group (no training or alternative training, such as maths). Participants were English-speaking children, adolescents or adults whose word reading was below the level expected for their age for no known reason (i.e., they had adequate attention and no known physical, neurological or psychological problems). Data collection and analysis. Two review authors independently selected studies, assessed risk of bias and	−0.46 to 0.74; three studies), spelling SMD 0.36 (95% CI −0.27 to 1.00; two studies), letter-sound knowledge SMD 0.35 (95% CI 0.04 to 0.65; three studies), and phonological output SMD 0.38 (95% −0.04 to 0.80; four studies). There was one result in a negative direction for non-word reading fluency SMD 0.38 (95% CI −0.55 to 1.32; one study), though this was not statistically significant. We did five subgroup analyses on two outcomes that had sufficient data (word reading accuracy and non-word reading accuracy). The efficacy of phonics training was not moderated significantly by training type (phonics alone versus phonics and phoneme awareness versus phonics and irregular word training), training intensity (less than two hours per week versus at least two hours per week), training duration (less than three months versus at least three months), training group size (one-on-one versus small group training), or training administrator (human administration versus computer administration). Authors' conclusions: Phonics training appears to be effective for improving some reading skills. Specifically, statistically significant effects were found for non-word reading accuracy (large effect), word reading accuracy (moderate effect), and letter-sound knowledge (small-to-moderate effect). For several other

extracted data. Main results: We found 11 studies that met the criteria for this review. They involved 736 participants. We measured the effect of phonics training on eight outcomes.

outcomes, there were small or moderate effect sizes that did not reach statistical significance but may be meaningful: word reading fluency, spelling, phonological output and reading comprehension. The effect for non-word reading fluency, which was measured in only one study, was in a negative direction, but this was not statistically significant.

Future studies of phonics training need to improve the reporting of procedures used for random sequence generation, allocation concealment and blinding of participants, personnel, and outcome assessment.

Sherman (2007)   0.39

The purpose of this study was to synthesise, using meta-analytical methods, the research regarding phonemic awareness and phonics (decoding) instruction with students in grades 5 through 12 who read significantly below grade level expectations. Twenty-six studies published between 1975 and 2005 met the criteria for inclusion and analysis. A total of 1,358 students participated in the studies (565 in control groups, 799 in treatment groups). The effect sizes of interventions = impact on achievement were calculated on five levels of dependent variables (word identification or word attack skills of sub-syllabic or single syllable levels, and decoding multi-syllabic words; oral reading fluency and accuracy of individual words or connected text; comprehending words or vocabulary; comprehending text; decoding, fluency and comprehension). Four separate analyses were

Oral reading d = 0.27 SD = 0.30, SE = 0.15 CI [−0.68, 1.21]

Text comprehension d = 0.42 SD = 0.28, SE = 0.15 CI [0.32, 0.52]

26 studies, 1,358 students, 88 ESs

9 to 18 years

Because of the small number of studies and the variability of the population studied, the alpha level was relaxed to 0.25 to explore statistical significance of main effects or interaction effects at this level. The impact of group size and reading level on effect size was significant in many of the analyses at a 0.25 alpha level.

Appendix A.1.1: (*cont.*)

Citation	Overall ES	Abstract	Moderator variables
		presented: (a) the full data set; (b) the data set with outliers removed; (c) the full data set without one study (Mercer et al., 2000); and (d) the data without outliers and without the Mercer study. Although many of the studies exhibited medium to high effect sizes, none of the analyses at an alpha level of 0.05 reached statistical significance. The results were mixed for group size/intervention focus and reading level/intervention focus. Significant main effects were found for reading level (reading level*intervention focus) and the interaction between group size and intervention focus on word identification or word attack skills of sub-syllabic or single syllable levels, and decoding multi-syllabic words. The impact of reading level, group size and intervention focus on effect size were not significant at any level. Limitations of this meta-analysis, features of interventions that show promise in accelerating the reading skills of delayed older readers and suggestions for future research are also presented.	
Slavin et al. (2011)	0.62 (one-to-one) 0.32 (small groups)	This article reviews research on the achievement outcomes of alternative approaches for struggling readers ages 5–10 (US grades K–5): One-to-one tutoring, small-group tutorials, classroom instructional process approaches and computer-assisted instruction. Study inclusion criteria included use of randomised	

or well-matched control groups, study duration of at least 12 weeks and use of valid measures independent of treatments. A total of 97 studies met these criteria. The review concludes that one-to-one tutoring is very effective in improving reading performance. Tutoring models that focus on phonics obtain much better outcomes than others. Teachers are more effective than paraprofessionals and volunteers as tutors. Small-group, phonetic tutorials can be effective but are not as effective as one-to-one phonetically focused tutoring. Classroom instructional process programmes, especially cooperative learning, can have very positive effects for struggling readers. Computer-assisted instruction had few effects on reading. Taken together, the findings support a strong focus on improving classroom instruction and then providing one-to-one, phonetic tutoring to students who continue to experience difficulties.

Torgerson et al. (2006)	0.27 SE 0.09	Executive summary: The Department for Education and Skills (DfES) commissioned the Universities of York and Sheffield to conduct a systematic review of experimental research on the use of phonics instruction in the teaching of reading and spelling. This review is based on evidence from randomised controlled trials (RCTs). Key findings: The effect of phonics on reading: Systematic phonics instruction within a broad literacy curriculum was found to have a	Systematic phonics instruction vs. whole language or whole word intervention (comprehension): 0.24. Systematic phonics instruction vs. whole language or whole word intervention (spelling): 0.09. Systematic synthetic phonics instruction vs. systematic analytic phonics instruction: 0.02. **Sample:** normal achieving children = 0.45, children with reading difficulties = 0.21.

Appendix A.1.1: (*cont.*)

Citation	Overall ES	Abstract	Moderator variables
		statistically significant positive effect on reading accuracy. There was no statistically significant difference between the effectiveness of systematic phonics instruction for reading accuracy for normally developing children and for children at risk of reading failure. The weight of evidence for both these findings was moderate (there were 12 randomised controlled trials included in the analysis). Both of these findings provided some support for those of a systematic review published in the United States in 2001 (Ehri et al., 2001). An analysis of the effect of systematic phonics instruction on reading comprehension was based on weak weight of evidence (only four randomised controlled trials were found) and failed to find the statistically significant positive difference which was found in the previous review. The effect of synthetic and analytic phonics: The weight of evidence on this question was weak (only three randomised controlled trials were found). No statistically significant difference in effectiveness was found between synthetic phonics instruction and analytic phonics instruction. The effect of phonics on spelling: The weight of evidence on this question was weak (only three randomised controlled trials were found). No effect of systematic phonics instruction on spelling was found.	

Appendix A.5.1: *Feedback meta-analyses*

Citation	Overall ES	Abstract	Moderator variables
Bangert-Drowns et al. (1991)	0.26 SE 0.06	Feedback is an essential construct for many theories of learning and instruction and an understanding of the conditions for effective feedback should facilitate both theoretical development and instructional practice. In an early review of feedback effects in written instruction Kulhavy (1977) proposed that feedback's chief instructional significance is to correct errors. This error-correcting action was thought to be a function of presentation timing, response certainty and whether students could merely copy answers from feedback without having to generate their own. The present meta-analysis reviewed 58 effect sizes from 40 reports. Feedback effects were found to vary with for control for pre-search availability, type of feedback, use of pretests and type of instruction and could be quite large under optimal conditions. Mediated intentional feedback for retrieval and application of specific knowledge appears to stimulate the correction of erroneous responses in situations where its mindful (Solomon and Globerson, 1987) reception is encouraged.	Type of feedback: explain and repeat until corrected has the largest effect size with right/wrong feedback the lowest. Timing of feedback: immediately after the largest effects. Larger effect sizes if it counts for grade and if it is multiple choice type. Larger effect sizes if it counts for grade and if it is multiple choice type. If there is control for presearch availability effect sizes are larger. Familiarity of criterion produces large effects when children are familiar than when they are not. Type of instruction: testing produces largest effects. All the above moderators are significant. Grade and subject matter have no significant differences of effects.
Fuchs and Fuchs (1986)	0.72 SE 0.09	Whilst the aptitude treatment interaction (ATI) approach to educational measurement emphasises establishing salient learner characteristics, systematic formative evaluation provides ongoing evaluation for instructional programme modification. Systematic formative evaluation appears more tenable than ATI	Publication type = journal largest effects. Year = more recent (compared to publication date) largest effects. Quality = good, data display = graphed, data evaluation= by rule yielded largest effects. Grade = 7–12-largest.

Appendix A.5.1: (*cont.*)

Citation	Overall ES	Abstract	Moderator variables
		for developing individualised instructional programmes. This meta-analysis investigates the effects of educational programmes on student achievement. Twenty-one controlled studies generated 95 relevant effect sizes, with an average effect size of 0.72. The magnitude of effect size was associated with publication type, data evaluation methods, and use of behaviour modification. Findings indicate that unlike reported ATI approaches to individualisation, systematic formative evaluation procedures reliably increase academic achievement. This suggests that, given an adequate measurement methodology, practitioners can inductively formulate successful individualised educational programmes.	Measurement frequency = twice a week-largest. Duration = more than 10 weeks-largest.
Graham et al. (2015)	0.61 CI 0.42 to 0.79	To determine whether formative writing assessments that are directly tied to everyday classroom teaching and learning enhance students' writing performance, we conducted a meta-analysis of true and quasi-experiments conducted with students in grades 1 to 8. We found that feedback to students about writing from adults, peers, self and computers statistically enhanced writing quality, yielding average weighted effect sizes of 0.87, 0.58, 0.62 and 0.38, respectively. We did not find, however, that teachers' monitoring of students' writing progress or implementation of the 6 _ 1 Trait Writing model meaningfully enhanced	Adult feedback = 0.87, peer = 0.58, self-assessment = 0.62, computer assessed = 0.38 27 studies = overall feedback d = 0.61 CI 0.42 to 0.79 *ages 5 to 14 *outcome: writing

		students' writing. The findings from this meta-analysis provide support for the use of formative writing assessments that provide feedback directly to students as part of everyday teaching and learning. We argue that such assessments should be used more frequently by teachers and that they should play a stronger role in the Next-Generation Assessment Systems being developed by Smarter Balanced and PARCC.	Moderator analyses suggested that formative assessment might be more effective in English language arts (ELA) than in mathematics or science, with estimated effect sizes of 0.32, 0.17 and 0.09, respectively. Two types of implementation of formative assessment, one based on professional development and the other on the use of computer-based formative systems, appeared to be more effective than other approaches, yielding mean effect size of 0.30 and 0.28, respectively.
Kingston and Nash (2011)	0.20 CI 0.19 to 0.21	An effect size of about 0.70 (or 0.40–0.70) is often claimed for the efficacy of formative assessment but is not supported by the existing research base. More than 300 studies that appeared to address the efficacy of formative assessment in grades K–12 were reviewed. Many of the studies had severely flawed research designs yielding un-interpretable results. Only 13 of the studies provided sufficient information to calculate relevant effect sizes. A total of 42 independent effect sizes were available. The median observed effect size was 0.25. Using a random effects model, a weighted mean effect size of 0.20 was calculated. Moderator analyses suggested that formative assessment might be more effective in English language arts (ELA) than in mathematics or science, with estimated effect sizes of 0.32, 0.17, and 0.09, respectively. Two types of implementation of formative assessment, one based on professional development and the other on the use of computer-based formative systems, appeared to be more effective than other approaches, yielding mean effect size of 0.30 and 0.28, respectively. Given the wide use	

Appendix A.5.1: (cont.)

Citation	Overall ES	Abstract	Moderator variables
		and potential efficacy of good formative assessment practices, the paucity of the current research base is problematic. A call for more high-quality studies is issued.	
Kluger and DeNisi (1996)	0.41 SE 0.09	Since the beginning of the century, feedback interventions (FIs) produced negative – but largely ignored – effects on performance. A meta-analysis (607 effect sizes; 23,663 observations) suggests that FIs improved performance on average ($d = 0.41$) but that over 1/3 of the FIs decreased performance. This finding cannot be explained by sampling error, feedback sign or existing theories. The authors proposed a preliminary FI theory (FIT) and tested it with moderator analyses. The central assumption of FIT is that FIs change the locus of attention among 3 general and hierarchically organised levels of control: task learning, task motivation and meta-tasks (including self-related) processes. The results suggest that FI effectiveness decreases as attention moves up the hierarchy closer to the self and away from the task. These findings are further moderated by task characteristics that are still poorly understood.	The moderator analyses suggest two major conclusions. First, several FI cues that seem to direct attention to meta-task processes *attenuate* FI effects on performance, whereas several FI cues that seem to direct attention to task-motivation or task-learning processes *augment* FI effects on performance. The second major conclusion is that FI effects are moderated by the nature of the task. However, the exact task properties that moderate FI effects are still poorly understood. First, simple-task performance benefited from FIs (marginally) more than complex-task performance. Second, the performance of novel tasks seemed to be debilitated when performance was measured for a short time (i.e., performance in the initial stages of task acquisition). Although the moderating effects of task features identified by FIT received weak support, several task dimensions moderated FI effects unexpectedly: Physical tasks and following rules tasks yielded weaker FI effects, and memory tasks yielded stronger FI effects.

| Lysakowski and Walberg (1982) | 0.97 SD 1.53 | To estimate the instructional effects of cues, participation and corrective feedback on learning 94 effect sizes were calculated from statistical data in 54 studies containing a combined sample of 14,689 students in approximately 700 classes. The mean of the study-weighted effect size is 0.97, which suggest average percentiles on learning outcomes of 83 and 50, respectively, for experimental and control groups. The strong effects appeared constant from elementary level through college and across socioeconomic levels, races, private and public schools, and community types. In addition the effects were not significantly different across the categories of methodological rigour such as experiments and quasi-experiments. |
| Tenenbaum and Goldring (1989) | 0.72 SE 0.37 | Estimated the effect of enhanced instruction on motor skill acquisition in a meta-analysis of 15 studies that used 4–5-yr-old children and 4th–21th graders in Israel. Students exposed to enhanced instruction gained more qualified motor skills than over 75% of the students exposed to regular instruction in a variety of motor skills. Enhanced instruction used cues and explanations by the instructor to clarify the motor skill, encouraged students to actively participate in the task over 70% of the time, reinforced students' responses and supplied ongoing feedback and correctives to ensure motor skill acquisition. |

Appendix A.5.2: *Metacognition and self-regulation meta-analyses*

Citation	Overall ES	Abstract	Moderator variables
Abrami et al. (2008)	0.34 SE 0.01 CI 0.31 to 0.37	Critical thinking (CT), or the ability to engage in purposeful, self-regulatory judgement, is widely recognised as an important, even essential, skill. This article describes an on-going meta-analysis that summarises the available empirical evidence on the impact of instruction on the development and enhancement of critical thinking skills and dispositions. We found 117 studies based on 20,698 participants, which yielded 161 effects with an average effect size (g+) of 0.341 and a standard deviation of 0.610. The distribution was highly heterogeneous (QT = 1,767.86, p<0.001). There was, however, little variation due to research design, so we neither separated studies according to their methodological quality nor used any statistical adjustment for the corresponding effect sizes. Type of CT intervention and pedagogical grounding were substantially related to fluctuations in CT effects sizes, together accounting for 32% of the variance. These findings make it clear that improvement in students' CT skills and dispositions cannot be a matter of implicit expectation. As important as the development of CT skills is considered to be, educators must take steps to make CT objectives explicit in courses and also to include them in both pre-service and in-service training and faculty development.	Publication bias = 0.10. Design type: pre-experimental = 0.31, quasi-experimental = 0.36, true-experimental = 0.34. Type of measure: standardised = 0.24, teacher made = 1.43, researcher made = 0.64, teacher/researcher = 0.16, secondary source = 0.29. Participant's age: elementary = 0.52, secondary = 0.69, high school = 0.10, undergraduate post secondary = 0.25, graduate post secondary = 0.62, adults = 0.32. Type of intervention: general critical thinking skills = 0.38, Infusion = 0.54, Immersion = 0.09, mixed = 0.94. Pedagogical grounding of intervention: Instructor training = 1.00, extensive observations = 0.58, detailed curriculum description = 0.31, critical thinking among course objectives = 0.13. Student collaboration: yes = 0.41, no = 0.31.

Chiu (1998)	0.67	In this paper, meta-analysis is used to identify components that are associated with effective metacognitive training programmes in reading research. Forty-three studies, with an average of 81 students per study, were synthesised. It was found that metacognitive training could be more effectively implemented by using small-group instruction, as opposed to large-group instruction or one-to-one instruction. Less intensive programmes were more effective than intensive programmes. Programme intensity was defined as the average number of days in a week that instruction was provided to students. Students in higher grades were more receptive to the intervention. Measurement artefacts, namely teaching to the test and use of non-standardised tests and the quality of the studies synthesised played a significant role in the evaluation of the effectiveness of the metacognitive reading intervention.	Metacognitive reading intervention = 0.24 Delivery of intervention: researcher or collaborators produced higher ES than teacher instruction. Intensity: less intensive programmes produced higher ES than high intensity ones. Setting: small group had the highest ES. Grade: fifth grade or higher produced highest ES. Students: low-ability highest ES (although marginally significant over other groups).
De Boer et al. (2014)	**reading** g = 0.36 SE = 0.08 **writing** g = 1.25 SE = 1.12 **math** g = 0.66 SE = 0.06 **science** g = 0.73 SE = 0.13	This meta-analysis examined the influence of attributes related to the implementation of learning strategy instruction interventions on students' academic performance and also examined how the attributes related to the method of testing the intervention effects affected the actual effects measured. Using meta-regression, we analysed the influence of the subject domain in which the intervention was implemented, the implementer, its duration and intensity, student cooperation, and research method aspects (including measurement instrument). Most attributes moderated the intervention effect. Using forward regression analysis, we only needed four attributes to obtain the best model, however. This	Implementer: researcher = 0.93, teacher = 0.60, PC = 0.55 Implementation fidelity: fidelity = 0.71, fidelity no but = 0.58, fidelity no = 0.61 Random assignment: yes = 0.70, no = 0.58 Control group: business as usual = 0.61, different situation = 0.72 Measurement instrument: self-developed = 0.25, standardised tests = 0.43

Appendix A.5.2: (cont.)

Citation	Overall ES	Abstract	Moderator variables
		analysis showed that the intervention effect was lower when a standardised test was used for evaluation instead of an unstandardised test. Interventions implemented by assistants or researchers were more effective than those implemented by teachers or using computers. Cooperation had a negative, and session duration a positive, contribution. Together, these attributes explained 63.2% of the variance in effect, which stresses the importance of emphasising not only the instructional focus of an intervention but also its other attributes.	
Dignath et al. (2008)	0.62 SE 0.05	Recently, research has increasingly focused on fostering self-regulated learning among young children. To consider this trend, this article presents the results of a differentiated meta-analysis of 48 treatment comparisons resulting from 30 articles on enhancing self-regulated learning among primary school students. Based on recent models of self-regulated learning, which consider motivational, as well as cognitive, and metacognitive aspects [Boekaerts, M. (1999). 'Self-Regulated Learning: Where We Are Today'. *International Journal of Educational Research*, 31(6), 445–457], the effects of self-regulated learning on academic achievement and on cognitive and metacognitive strategy application, as well as on motivation were analysed. As the results show, self-regulated learning training programmes proved to be effective, even at primary school level. Subsequent analysis tested for the effects of several moderator	School subject: maths = 0.82, reading/writing = 0.55, other = 0.49. Theoretical background: metacognitive = 0.58, motivational = 0.33, social-cognitive = 0.95 Instructed cognitive strategy: no cognitive strategy = 0.63, elaboration = 0.84, problem-solving = 0.98, elaboration and organisation = 0.37, elaboration and problem-solving = 0.42. Instructed metacognitive strategy: no metacognitive = 0.66, planning = 0.60, monitoring = 0.91, evaluation = 0.69, planning and monitoring = 0.78, monitoring and evaluation = 0.35, all three = 0.56. Instructed metacognitive reflection: no metacognitive= 0.60, reasoning = 0.83, benefit of strategy use = 0.61, reasoning and knowledge = 0.35, reasoning and benefit = 0.66, knowledge and benefit = 0.55, all three = 0.73.

variables, which consisted of study features and training characteristics. Regarding factors that concern the content of the treatment and the impact of the theoretical background that underlies the intervention was tested, as well as the type of cognitive, metacognitive or motivational strategy which were instructed, and if group work was used as instruction method. Training-context-related factors, which were included in the analyses, consisted of students' grade level, the length of the training, and if teachers or researchers directed the intervention, as well as the school subject in which context the training took place. Following the results of these analyses, a list with the most effective training characteristics was provided.

Instructed motivational strategy: no motivational strategy = 0.58, causal attribution = 0.64, action control = 0.49, feedback = 1.41, feedback and causal attribution = 0.33.
Instruction method: group work = 0.74, no group work = 0.50.
Direction of training: researcher-directed = 0.87, teacher-directed = 0.46.
Type of assessment: task = 0.69, multiple choice = 0.35, simulation task = 0.55, others = 0.57.

Donker et al. (2014)

0.66
SE = 0.05
CI 0.56 to 0.76

In this meta-analysis the results of studies on learning strategy instruction focused on improving self-regulated learning were brought together to determine which specific strategies were the most effective in increasing academic performance. The meta-analysis included 58 studies in primary and secondary education on interventions aimed at improving cognitive, metacognitive and management strategy skills, as well as motivational aspects and metacognitive knowledge. A total of 95 interventions and 180 effect sizes demonstrated substantial effects in the domains of writing (Hedges' g = 1.25), science (0.73), mathematics (0.66) and comprehensive reading (0.36). These domains differed in terms of which strategies were the most effective in improving academic performance. However, metacognitive knowledge instruction appeared to be valuable in all of them. Furthermore, it was found that the effects were higher when self-

Cognitive strategies:
rehearsal =1.39
elaboration = 0.75
Organisation = 0.81
Metacognitive strategies:
planning = 0.80
monitoring = 0.71
evaluation = 0.75
Metacognitive knowledge:
general = 0.97
personal = 0.94
Student characteristics:
regular = 0.61
low SES and ethnic minority = 0.72
special needs = 0.89
gifted and high SES = 0.72
Measurement instruments:
self-developed = 0.78

Appendix A.5.2: (*cont.*)

Citation	Overall ES	Abstract	Moderator variables
		developed tests were used than in the case of intervention-independent tests. Finally, no differential effects were observed for students with different ability levels. To conclude, the authors have listed some implications of their analysis for the educational practice and made some suggestions for further research.	intervention independent = 0.45 School subject: writing = 1.25 science = 0.73 math = 0.66 comprehensive reading = 0.36
Haller et al. (1988)	0.71	To assess the effect of 'metacognitive' instruction on reading comprehension, 20 studies, with a total student population of 1,553, were compiled and quantitatively synthesised. For 115 effect sizes or contrasts of experimental and control groups' performance, the mean effect size was 0.71, which indicates a substantial effect. In this compilation of studies, metacognitive instruction was found particularly effective for junior high students (seventh and eighth grades). Among the metacognitive skills, awareness of textual inconsistency and the use of self-questioning and the use of self-questioning and a monitoring and a regulating strategy were most effective. Reinforcement was the most effective teaching strategy.	Grade: Seventh and eighth grade largest ES. Instruction length: 10 minutes or less is insufficient. Metacognitive activities: Textual-dissonance approach in awareness, self-questioning strategy in monitoring, backward-forward and self-questioning approaches in regulating were the most effective. Instructional approach: reinforcement produced statistically significant effects.
Higgins et al. (2005)	0.62 SE 0.09	Executive summary: Relevant studies in the area of thinking skills were obtained by systematically searching a number of online databases of educational research literature, by identifying references in reviews and other relevant books and reports, and from contacts with expertise in this area. Twenty-six of the studies identified for this review were obtained from the database which resulted from the first thinking skills review (Higgins et al., 2004); a further three resulted from updating the original search and	Weight of evidence: high = 0.57, medium = 0.77, low = 0.43. Programme type: instrumental enrichment = 0.58, cognitive acceleration = 0.61, metacognitive strategies = 0.96. Curriculum areas: reading = 0.48, mathematics = 0.89, science = 0.78.

applying the more stringent criteria required for a quantitative synthesis. Studies were selected for the meta-analysis if they had sufficient quantitative data to calculate an effect size (relative to a control or comparison group of pupils) and if the number of research subjects was greater than 10. Effect sizes were calculated from the reported data and combined statistically using quantitative synthesis. Results: 29 studies were identified which contained quantitative data on pupils' attainment and attitudes suitable for meta-analysis. The studies come from a range of countries around the world with half set in the US or UK. The studies broadly cover the ages of compulsory schooling (5–16) and include studies set in both primary and secondary schools. A number of named thinking skills interventions are included, such as Feuerstein's instrumental enrichment (FIE) and cognitive acceleration through science education (CASE) as well as studies which report a more general thinking skills approach (such as the development of metacognitive strategies). The quantitative synthesis indicates that thinking skills programmes and approaches are effective in improving the performance on tests of cognitive measures (such as Raven's progressive matrices) with an overall effect size of 0.62. (This effect would move a class ranked at 50th place in a league table of 100 similar classes to 26th or a percentile gain of 24 points.) However, these approaches also have a considerable impact on curricular outcomes with the same effect size of 0.62. The overall effect size (including cognitive, curricular

Appendix A.5.2: (*cont.*)

Citation	Overall ES	Abstract	Moderator variables

and affective measures) was 0.74. Conclusions: Overall, the quantitative synthesis indicates that, when thinking skills programmes and approaches are used in schools, they are effective in improving pupils' performance on a range of tested outcomes (relative to those who did not receive thinking skills interventions). The magnitude of the gains found appears to be important when compared with the reported effect sizes of other educational interventions. This review found an overall mean effect of 0.62 for the main (cognitive) effect of each of the included studies, larger than the mean of Hattie's vast database of meta-analyses at 0.40 (Hattie, 1999) but very similar to the overall figure reported by Marzano (1998: p 76) of 0.65 for interventions across the knowledge, cognitive, metacognitive and self-system domains. In particular, our study identified metacognitive interventions as having relatively greater impact, similar to Marzano's study. Looking at a smaller part of our review, Feuerstein's instrumental enrichment is one of the most extensively researched thinking skills programme. Our results broadly concur with those of Romney and Samuels (2001), whose meta-analysis found moderate overall effects and an effect size of 0.43 on reasoning ability. Our findings were of the same order, with an overall effect size of 0.58 (one main effect from each of seven studies included) and an effect size of 0.52 on tests of reasoning (one main effect from four studies). There is

Study	Effect size	Description	Notes
		some indication that the impact of thinking skills programmes and approaches may vary according to subject. In our analysis, there was relatively greater impact on tests of mathematics (0.89) and science (0.78), compared with reading (0.40).	
Klauer and Phye (2008)	0.69 SE 0.05	Researchers have examined inductive reasoning to identify different cognitive processes when participants deal with inductive problems. This article presents a prescriptive theory of inductive reasoning that identifies cognitive processing using a procedural strategy for making comparisons. It is hypothesised that training in the use of the procedural inductive reasoning strategy will improve cognitive functioning in terms of (a) increased fluid intelligence performance and (b) better academic learning of classroom subject matter. The review and meta-analysis summarises the results of 74 training experiments with nearly 3,600 children. Both hypotheses are confirmed. Further, two moderating effects were observed: Training effects on intelligence test performance increased over time, and positive problem-solving transfer to academic learning is greater than transfer to intelligence test performance. The results cannot be explained by placebo or test-coaching effects. It is concluded that the proposed strategy is theoretically and educationally promising and that children of a broad age range and intellectual capacity benefit with such training.	Programme type: Programme I (kindergarten/primary) = 0.64, Programme II (secondary) = 0.64, Programme III (older) = 0.84. Subjects: primary = 0.63, secondary = 0.59, special education = 0.94. Training conditions: small groups = 0.73, classes = 0.62.
Zheng (2016)	0.44 SE = 0.10	This meta-analysis examined research on the effects of self-regulated learning scaffolds on academic	29 studies, 2,648 students, K–12 (5 to 18 years old)

Appendix A.5.2: (*cont.*)

Citation	Overall ES	Abstract	Moderator variables
	CI 0.22 to 0.64	performance in computer-based learning environments from 2004 to 2015. A total of 29 articles met inclusion criteria and were included in the final analysis with a total sample size of 2,648 students. Moderator analyses were performed using a random effects model that focused on the three main areas of scaffold characteristics (including the mechanism, functions, delivery forms, mode and number of scaffolds; how to promote self-regulated learning by scaffolds); demographics of the selected studies (including sample groups, sample size, learning domain, research settings and types of computer-based learning environments); and research methodological features (including research methods, types of research design, types of organisation for treatment and duration of treatment). Findings revealed that self-regulated learning scaffolds in computer-based learning environments generally produced a significantly positive effect on academic performance (ES = 0.438). It is also suggested that both domain-general and domain-specific scaffolds can support the entire process of self-regulated learning since they demonstrated substantial effects on academic performance. Different impacts of various studies and their methodological features are presented and discussed.	Mechanisms of scaffolds: prompts of hints = 0.31, concept map = 0.37, worked-out example = −0.03, integrated SRL tool = 0.94 Scaffolds functions: conceptual = 0.06, strategic = 1.04, metacognitive = 0.25, multiple = 0.37 Delivery method: direct = 0.35, indirect = 0.41, both = 1.20 Delivery modes: domain-general = 0.43, domain-specific = 0.10, both = 1.69 Number of scaffolds: unique = 0.21, multiple = 0.57 Sample groups: primary = 0.27, junior and senior = 0.84, higher education = 0.28 Sample size: 20–50 = 0.25, 51–100 = 0.36, 101–300 = 0.63 Learning domain: natural science = 0.45, social science = 0.41, medical = 0.27 Research settings: laboratory = 0.26, classroom = 0.75 Research design: true experiment = 0.41, Quasi-experimental = 0.48 Duration: less than one hour = 0.09, 1–10 h = 0.35, 1–7 days = 0.44, 2–4 weeks = 0.77, 5–8 weeks = 0.56, 9–24 weeks = −0.03

Appendix A.5.3: *Digital technology meta-analyses*

Title	Overall ES	Abstract	Moderator variables
Bayraktar (2001)	0.27	This meta-analysis investigated how effective computer-assisted instruction (CAI) is on student achievement in secondary and college science education when compared to traditional instruction. An overall effect size of 0.273 was calculated from 42 studies yielding 108 effect sizes, suggesting that a typical student moved from the 50th percentile to the 62nd percentile in science when CAI was used. The results of the study also indicated that some study characteristics such as student-to-computer ratio, CAI mode and duration of treatment were significantly related to the effectiveness of CAI.	The results of this analysis also indicated that all variables except educational level were related to effect size. The strongest relationships were found for the following variables: length of treatment, student-to-computer ratio, and publication year. Effect sizes did not vary by publication status and educational level. This study detected a significant relationship between CAI effectiveness and instructional role of computers. Effect sizes were higher (ES = 0.288) when computers were used as a supplement to the regular instruction and lower when the computer entirely replaced the regular instruction, (ES = 0.178). This finding was consistent with the previous meta-analyses (Kulik et al., 1983; Liao, 1998) suggesting that using the computer as a supplement to regular instruction should be the preferred choice instead of using it as a replacement. This meta-analysis indicated that there were no significant effect size differences in different school levels. This result supports the meta-analysis conducted by Flinn and Gravat (1995) reporting an effect size of 0.26 standard deviations for elementary grades, an effect size of 0.20 standard deviations for secondary

Appendix A.5.3: (*cont.*)

Title	Overall ES	Abstract	Moderator variables
			grades, and an effect size of 0.20 standard deviations for college. However, this finding is not consistent with the majority of meta-analyses (Bangert-Drowns, 1985; Burns and Bozeman, 1981; Liao, 1998; Roblyer, 1989) that report significant effect size differences for different school levels.
			The results of this study indicated that the length of the treatment was strongly related to the effectiveness of CAI for teaching science. CAI was especially effective when the duration of treatment was limited to four weeks or less. The average effect of CAI in such studies was 0.404 standard deviations. In studies where treatment continued longer than four weeks, the effects were less clear (ES = 0.218). A similar relationship between length of treatment and study outcome has been reported in previous meta-analyses. Kulik et al. (1983), for example, reported an effect size of 0.56 for four weeks or less, 0.30 for 5–8 weeks, and 0.20 for more than eight weeks.
			This study concluded that the results found in ERIC documents were more positive (ES = 0.337) than results found in journal articles (ES = 0.293) and dissertations (ES = 0.229).

| Cheung and Slavin (2013) | 0.15 | Random effects | The present review examines research on the effects of educational technology applications on mathematics achievement in K–12 classrooms. Unlike previous reviews, this review applies consistent inclusion standards to focus on studies that met high methodological standards. In addition, methodological and substantive features of the studies are investigated to examine the relationship between educational technology applications and study features. A total of 74 qualified studies were included in our final analysis with a total sample size of 56,886 K–12 students: 45 elementary studies ($N = 31,555$) and 29 secondary studies ($N = 25,331$). Consistent with the more recent reviews, the findings suggest that educational technology applications generally produced a positive, though modest, effect ($ES = +0.15$) in comparison to traditional methods. However, the effects may vary by educational technology type. Among the three types of educational technology applications, supplemental CAI had the largest effect with an effect size of $+0.18$. The other two interventions, computer-management learning and comprehensive programmes, had a much smaller effect size, $+0.08$ and $+0.07$, respectively. Differential impacts by various study and methodological features are also discussed. | A marginally significant between-group effect ($QB = 5.58$, $df = 2$, $p < 0.06$) was found, indicating some variation among the three programmes. The 37 supplemental technology programmes produced the largest effect size, $+0.18$, and the seven computer-managed learning programmes and the eight comprehensive models produced similar small effect sizes of $+0.08$ and $+0.06$, respectively. The effect sizes for low, medium, and high intensity were $+0.03$, $+0.20$, and $+0.13$, respectively. In general, programmes that were used more than 30 minutes a week had a bigger effect than those that were used less than 30 minutes a week. The average effect size of studies with a high level of implementation ($ES = +0.26$) was significantly greater than those of low and medium levels of implementation ($ES = +0.12$). The effect sizes for low and high SES were $+0.12$ and $+0.23$, respectively. The difference between elementary studies ($ES = +0.17$) and secondary studies ($ES = +0.13$) was not statistically different. No publication bias was found No trend towards more positive results in recent years. The mean effect sizes for studies in the 1980s, 1990s and after 2000 were $+0.23$, $+0.15$ and $+0.12$, respectively. |

Appendix A.5.3: (*cont.*)

Title	Overall ES	Abstract	Moderator variables
			The mean effect size for quasi-experimental studies was +0.19, twice the size of that for randomised studies (+0.10). The mean effect size for the 30 small studies (ES = +0.26) was about twice that of large studies (ES = +0.12, p<0.01). Large randomised studies had an effect size of +0.08, whereas small randomised studies had an effect size that was twice as large (ES = +0.17).
Li and Ma (2010)	0.71	This study examines the impact of computer technology (CT) on mathematics education in K–12 classrooms through a systematic review of existing literature. A meta-analysis of 85 independent effect sizes extracted from 46 primary studies involving a total of 36,793 learners indicated statistically significant positive effects of CT on mathematics achievement. In addition, several characteristics of primary studies were identified as having effects. For example, CT showed advantage in promoting mathematics achievement of elementary over secondary school students. As well, CT showed larger effects on the mathematics achievement of special need students than that of general education students, the positive effect of CT	Four characteristics of the studies remained statistically significant collectively. Two of them indicated large effects. With other statistically significant variables controlled, special education status showed a magnitude of 1.02 SD in favour of applying technology to special need students over general education students, and method of teaching showed a magnitude of 0.79 SD in favour of using technology in school settings where teachers practiced constructivist approach to teaching over school settings where teachers practiced traditional approach to teaching. Meanwhile, two characteristics indicated moderate and small effects of technology on mathematics achievement. Year of publication showed a moderate magnitude of 0.32 SD in favour of

was greater when combined with a constructivist approach to teaching than with a traditional approach to teaching, and studies that used non-standardised tests as measures of mathematics achievement reported larger effects of CT than studies that used standardised tests. The weighted least squares univariate and multiple regression analyses indicated that mathematics achievement could be accounted for by a few technology, implementation and learner characteristics in the studies.

publications before the turn of the century (before 1999) over publications after the turn of the century (after 1999), with other statistically significant variables controlled. Type (or level) of education showed a small magnitude of 0.22 SD in favour of using technology at the elementary school level over second school level, with other statistically significant variables controlled.

Lou et al. (2001)   0.16

This study quantitatively synthesised the empirical research on the effects of social context (i.e., small group versus individual learning) when students learn using computer technology. In total, 486 independent findings were extracted from 122 studies involving 11,317 learners. The results indicate that, on average, small group learning had significantly more positive effects than individual learning on student individual achievement (mean ES = +0.15), group task performance (mean ES = +0.31), and several process and affective outcomes. However, findings on both individual achievement and group task performance were significantly heterogeneous. Through weighted least squares univariate and multiple regression analyses, we found that variability in each of the two cognitive outcomes could be accounted for by a few

The overall effect of social context on individual achievement was based on 178 independent effect sizes extracted from 100 studies. The mean weighted effect size ($d+$) was +0.16 (95% CI is +0.12 to +0.20; and $QT = 341.95$, $df = 177$, $p<.05$) before outlier procedures. Individual effect sizes ranged from −1.14 to +3.37, with 105 effect sizes above zero favouring learning in groups, 15 effect sizes equal to zero and 58 effect sizes below zero favouring individual learning. Fifteen outliers with standardised residuals larger than ±2.00 were identified. After outlier procedures, the mean effect size was +0.15 (95% confidence interval is +0.11 to +0.19). The results indicate that, on average, there was a small but significantly positive effect of small group learning on student achievement as measured

Appendix A.5.3: (*cont.*)

Title	Overall ES	Abstract	Moderator variables
		technology, task, grouping and learner characteristics in the studies. The results of hierarchical regression model development indicate that the effects of small group learning with CT on individual achievement were significantly larger when: (a) students had group work experience or specific instruction for group work rather than when no such experience or instruction was reported; (b) cooperative group learning strategies were employed rather than general encouragement only or individual learning strategies were employed; (c) programmes involved tutorials or practice or programming languages rather than exploratory environments or as tools for other tasks; (d) subjects involved social sciences or computer skills rather than mathematics, science, reading, and language arts; (e) students were relatively low in ability rather than medium or high in ability; and (f) studies were published in journals rather than not published. When all the positive conditions were present, students learning in small groups could achieve 0.66 standard deviation more than those learning individually. When none of the positive conditions were present, students learning individually could learn 0.20 standard deviation more than those learning in groups.	by individually administered immediate or delayed post-tests. Effect sizes were significantly larger when students were learning with tutor programmes ($d = +0.20$) or programming languages ($d = +0.22$) than when using exploratory or tool programmes ($d = +0.04$). Effect sizes greatest for low attaining learners (0.34) as compared with medium (0.09), high (0.24) or mixed (0.12). The effects of social context on student individual achievement were larger when the subjects involved were computer skills ($d+ = +0.24$), social sciences and other ($d+ = +0.20$) than when the subjects were math/science/language arts ($d = +0.11$). The effect sizes were significantly positive for both heterogeneous ability groups ($d = +0.21$) and homogeneous ability groups ($d = +0.22$). Effect sizes were significantly more positive when specific cooperative learning strategies were employed ($d = +0.21$) than when students were generally encouraged to work together ($d = -0.04$) or when students in groups worked under individualistic goals or when no group learning strategy was described in the study ($d = +0.08$), with the latter two means not significantly different from zero.

Significantly more positive when students worked in pairs (d = +0.18) than when they worked in three to five member groups (d = +0.08).

Type of feedback, types of tasks, task familiarity, task difficulty, number of sessions, session duration, grade level, gender, computer experience, instructional control and whether achievement outcomes measured were of higher-order skills or lower-order skills were not found to be significantly related to the variability in the effects of social context on student individual achievement. Individuals appear to benefit from computer-based feedback but groups do better without computer-based feedback when completing group tasks.

Means et al. (2009) 0.24

A systematic search of the research literature from 1996 through July 2008 identified more than 1,000 empirical studies of online learning. Analysts screened these studies to find those that (a) contrasted an online to a face-to-face condition, (b) measured student learning outcomes, (c) used a rigorous research design and (d) provided adequate information to calculate an effect size. As a result of this screening, 51 independent effects were identified that could be subjected to meta-analysis. The meta-analysis found that, on average, students in online learning conditions performed better than those receiving face-to-face instruction. The

Students who took all or part of their class online performed better, on average, than those taking the same course through traditional face-to-face instruction. Learning outcomes for students who engaged in online learning exceeded those of students receiving face-to-face instruction, with an average effect size of +0.24 favouring online conditions. The mean difference between online and face-to-face conditions across the 51 contrasts is statistically significant at the $p<0.01$ level. Interpretations of this result, however, should take into consideration the fact that online and face-to-face conditions generally differed on

Appendix A.5.3: *(cont.)*

Title	Overall ES	Abstract	Moderator variables
		difference between student outcomes for online and face-to-face classes – measured as the difference between treatment and control means, divided by the pooled standard deviation – was larger in those studies contrasting conditions that blended elements of online and face-to-face instruction with conditions taught entirely face-to-face. Analysts noted that these blended conditions often included additional learning time and instructional elements not received by students in control conditions. This finding suggests that the positive effects associated with blended learning should not be attributed to the media, per se. An unexpected finding was the small number of rigorous published studies contrasting online and face-to-face learning conditions for K–12 students. In light of this small corpus, caution is required in generalising to the K–12 population because the results are derived for the most part from studies in other settings (e.g., medical training, higher education).

Few rigorous research studies of the effectiveness of online learning for K–12 students have been published. A systematic search of the research literature from 1994 through 2006 found no experimental or controlled quasi-experimental studies comparing the learning effects of online | multiple dimensions, including the amount of time that learners spent on task. The advantages observed for online learning conditions therefore may be the product of aspects of those treatment conditions other than the instructional delivery medium per se. Instruction combining online and face-to-face elements had a larger advantage relative to purely face-to-face instruction than did purely online instruction. The mean effect size in studies comparing blended with face-to-face instruction was +0.35, $p<0.001$. This effect size is larger than that for studies comparing purely online and purely face-to-face conditions, which had an average effect size of +0.14, $p<0.05$. An important issue to keep in mind in reviewing these findings is that many studies did not attempt to equate (a) all the curriculum materials, (b) aspects of pedagogy and (c) learning time in the treatment and control conditions. Indeed, some authors asserted that it would be impossible to have done so. Hence, the observed advantage for online learning in general, and blended learning conditions in particular, is not necessarily rooted in the media used per se and may reflect differences in content, pedagogy and learning time. |

versus face-to-face instruction for K–12 students that provide sufficient data to compute an effect size. A subsequent search that expanded the time frame through July 2008 identified just five published studies meeting meta-analysis criteria.

Studies in which learners in the online condition spent more time on task (0.46) than students in the face-to-face condition found a greater benefit for online learning … compared with +0.19 for studies in which the learners in the face-to-face condition spent as much time or more on task

Effect sizes were larger (0.42) for studies in which the online and face-to-face conditions varied in terms of curriculum materials and aspects of instructional approach in addition to the medium of instruction rather than those which replicated the instruction and curriculum (0.20).

The meta-analysis did not find differences in average effect size between studies published before 2004 (which might have used less sophisticated Web-based technologies than those available since) and studies published from 2004 on (possibly reflecting the more sophisticated graphics and animations or more complex instructional designs available). Nor were differences associated with the nature of the subject matter involved.

Finally, the examination of the influence of study method variables found that effect sizes did not vary significantly with study sample size or with type of design.

Moran et al. (2008)   0.49

The results of a meta-analysis of 20 research articles containing 89 effect sizes related to the use of digital tools and learning environments

1. The effect sizes were greater for interventions aimed at general populations than those with specific needs (i.e., students who are learning disabled or struggling readers). For the 57

Appendix A.5.3: (cont.)

Title	Overall ES	Abstract	Moderator variables
		to enhance literacy acquisition for middle school students demonstrate that technology can have a positive effect on reading comprehension (weighted effect size of 0.489). Very little research has focused on the effect of technology on other important aspects of reading, such as metacognitive, affective and dispositional outcomes. The evidence permits the conclusion that there is reason to be optimistic about using technology in middle school literacy programmes, but there is even greater reason to encourage the research community to redouble its efforts to investigate and understand the impact of digital learning environments on students in this age range and to broaden the scope of the interventions and outcomes studied.	effect sizes reported for a general, undifferentiated population of middle school students, the mean effect size was 0.52, whereas the effect size for targeted populations of students (e.g., students classified as possessing learning disabilities or as struggling readers) was 0.32: this was a statistically reliable difference. We can only speculate about why this might be the case, and we surely need more evidence before reaching a definitive conclusion. However, issues of engagement and appropriate levels of support and feedback suggest themselves as reasonable explanations. 2. Standardised measures from test companies (0.30), were less sensitive to treatment effects than researcher-developed measures in several of the studies in this meta-analysis measures were less sensitive to treatment effects than experimenter-designed assessments (0.56). 3. Sample size was a robust predictor of effect size; small n studies (30 or less) produced 14 effect sizes averaging 0.77, whilst large n (31 or more) studies produced 75 effect sizes with a mean of 0.38, $Q D 3:24; p< 0:20$. Studies with smaller sample sizes were much more likely to achieve substantial effects than those

with larger sample sizes. This counter-intuitive finding is puzzling because of what we know about the increase in statistical power that comes with larger experimental samples. On the other hand, there may be a trade-off between statistical power and experimental precision; that is, it may be easier for researchers to maintain a high degree of fidelity to treatment in smaller studies because of the greater manageability prospects.

4. Technologies that were created by a research team (1.20) had a much larger effect size than those technologies either adapted from the commercial market (0.28) or those that merely used the technology as a delivery system (0.34). This finding may be related to the fact that those technologies created by researchers tended to have a clear theoretical focus that was matched by the assessments employed by the team. In short, alignment between intention and outcome measure may be the operative variable behind this robust finding.

5. Studies that used some sort of correlated design (pre-tests used as covariates for post-test or repeated measures designs in which the same subjects cycle through different interventions) are more likely to find reliable differences between interventions than are independent group designs.

Appendix A.5.3: *(cont.)*

Title	Overall ES	Abstract	Moderator variables
			6. Effect sizes in studies lasting two to four weeks (0.55) were larger than those in studies lasting less than a week (0.48) but much larger than those from studies lasting five or more weeks (0.34).
Morphy and Graham (2012)	0.52 – writing quality 0.48 – length 0.66 – development/ organisation of text 0.61 mechanical correctness	Since its advent word processing has become a common writing tool, providing potential advantages over writing by hand. Word processors permit easy revision, produce legible characters quickly, and may provide additional supports (e.g., spellcheckers, speech recognition). Such advantages should remedy common difficulties among weaker writers/ readers in grades 1–12. Based on 27 studies with weaker writers, 20 of which were not considered in prior reviews, findings from this meta-analysis support this proposition. From 77 independent effects, the following average effects were greater than zero: writing quality (d = 0.52), length (d = 0.48), development/ organisation of text (d = 0.66), mechanical correctness (d = 0.61), motivation to write (d = 1.42), and preferring word processing over writing by hand (d = 0.64). Especially powerful writing quality effects were associated with word processing programmes that provided text quality feedback or prompted planning,	Whilst basic word processing impacted writing quality positively, neither the addition of external instructional support (WP+; Δ = −0.28; p = 0.123) nor the use of voice recognition (VR; Δ = −0.20; p = 0.26) differed significantly from basic word processors alone. Conversely, three interventions which added internal support to the word processor (WP+ +) were associated with considerable gains in writing quality (Δ = 0.91; p = 0.002) when compared to P&P. Random assignment and rater blinding were both associated with larger writing quality effects, but only random assignment proved statistically significant.

drafting, or revising (d = 1.46), although this observation was based on a limited number of studies (n = 3).

Onuoha (2007) 0.26

The purpose of this research study was to determine the overall effectiveness of computer-based laboratory compared with the traditional hands-on laboratory for improving students' science academic achievement and attitudes towards science subjects at the college and pre-college levels of education in the United States. Meta-analysis was used to synthesis the findings from 38 primary research studies conducted and/or reported in the United States between 1996 and 2006 that compared the effectiveness of computer-based laboratory with the traditional hands-on laboratory on measures related to science academic achievements and attitudes towards science subjects. The 38 primary research studies, with total subjects of 3,824 generated a total of 67 weighted individual effect sizes that were used in this meta-analysis. The study found that computer-based laboratory had small positive effect sizes over the traditional hands-on laboratory (ES = +0.26) on measures related to students' science academic achievements and attitudes towards science subjects (ES = +0.22). It was also found that computer-based laboratory produced more significant effects on physical science subjects

A total of 35 independent primary studies with a total of 3,284 subjects met the inclusion criteria to answer the primary research question. The 35 primary studies generated 35 weighted effect sizes (w), one effect size (d) for each primary study. The individual effect sizes ranged from a low negative effect size of −0.38, to a large positive effect size of +1.12. The overall mean effect size (ES), calculated at 95% confidence interval was +0.26 standard deviation units. Twelve primary studies representing 34% of the studies analysed reported negative effect sizes. The overall mean effect size (ES) for physical science subjects (physics and chemistry) was +0.34 standard deviation units, whilst biological science was +0.17 standard deviation units. The effect on pre-college studies was +0.24 standard deviation units compared to +0.21 obtained for college level studies.

ES for science attainment in studies between 1996 and 2000 +0.33 compared with +0.19 standard deviation units for studies conducted between 2001 and 2006.

Appendix A.5.3: (cont.)

Title	Overall ES	Abstract	Moderator variables
Rosen and Salomon (2007)	0.11	Different learning environments provide different learning experiences and ought to serve different achievement goals. We hypothesised that constructivist learning environments lead to the attainment of achievements that are consistent with the experiences that such settings provide and that more traditional settings lead to the attainments of other kinds of achievement in accordance with the experiences they provide. A meta-analytic study was carried out on 32 methodologically appropriate experiments in which these two settings were compared. Results supported one of our hypotheses showing that overall constructivist learning environments are more effective than traditional ones (ES = 0.460) and that their superiority increases when tested against constructivist-appropriate measures (ES = 0.902). However, contrary to expectations, traditional settings did not differ from constructivist ones when traditionally appropriate measures were used. A number of possible interpretations are offered among them the possibility that traditional settings have come to incorporate some constructivist	compared to biological sciences (ES = +0.34, +0.17). Grade level was found to moderately affect the result – although the effect sizes favoured CTILE regardless of grade level, still the effect size for grades 1–6 was significantly smaller than those for grades 7–9 (d+ = 0.413, d+ = 0.583 respectively, Qb = 5.29, p<0.05). Constructivist learning environments yielded significantly higher achievements than traditional ones when math instruction lasted for up to six weeks as compared with instruction that lasted for seven weeks or more (d+ = 0.686, d+ = 0.408, respectively, Qb = 10.76, p<0.01). Also year of publication made a difference – CTILE yielded larger effect sizes when the studies were published between 1986 and 1991 than between 1992 and 2002 (d+ = 0.554, d+ = 0.388, respectively, Qb = 6.16, p<0.05).

elements. This possibility is supported by other findings of ours such as smaller effect sizes for more recent studies and for longer lasting periods of instruction.

| Seo and Bryant (2009) | 0.37 | The purpose of this study was to conduct a meta-study of computer-assisted instruction (CAI) studies in mathematics for students with learning disabilities (LD) focusing on examining the effects of CAI on the mathematics performance of students with LD. This study examined a total of 11 mathematics CAI studies, which met the study selection criterion, for students with LD at the elementary and secondary levels and analysed them in terms of their comparability and effect sizes. Overall, this study found that those CAI studies did not show conclusive effectiveness with relatively large effect sizes. The methodological problems in the CAI studies limit an accurate validation of the CAI's effectiveness. Implications for future mathematics CAI studies were discussed. | CAI versus teacher instruction: The four group-design studies were associated with a small to medium effect size (d = 0.09, 0.33, 0.45 and 0.75). Comparison of CAI types: The two group-design studies compared the effectiveness of drill and practice CAI with game CAI for enhancing the addition skills of students with LD. Results of these studies demonstrated contradictory findings (d = 0.71 and −0.47 for drill and practice CAI). Enhanced CAI: The two group-design studies were related with either a small or large effect size (d = 0.87 and 0.30). The purpose of this study was to conduct a meta-study of mathematics CAI studies for students with LD. The 11 mathematics CAI studies were selected and examined their effectiveness for enhancing the mathematics performance of students with LD with their effect sizes. The results of this study found that the CAI studies in mathematics did not show conclusive effectiveness for the mathematics performance of students with LD with relatively large effect sizes. |

Appendix A.5.3: (*cont.*)

Title	Overall ES	Abstract	Moderator variables
Strong et al. (2011)		Fast ForWord is a suite of computer-based language intervention programmes designed to improve children's reading and oral language skills. The programmes are based on the hypothesis that oral language difficulties often arise from a rapid auditory temporal processing deficit that compromises the development of phonological representations. Methods: A systematic review was designed, undertaken and reported using items from the PRISMA statement. A literature search was conducted using the terms 'Fast ForWord' 'Fast For Word' 'Fastforword' with no restriction on dates of publication. Following screening of (a) titles and abstracts and (b) full papers, using pre-established inclusion and exclusion criteria, six papers were identified as meeting the criteria for inclusion (randomised controlled trial (RCT) or matched group comparison studies with baseline equivalence published in refereed journals). Data extraction and analyses were carried out on comparing the Fast ForWord intervention groups to both active and untreated control groups. Results: Meta-analyses indicated that there was no significant effect of Fast ForWord on any outcome measure in comparison to active or untreated control groups. Conclusions: There is	For the four analyses of Fast ForWord compared to untreated control groups, the pooled effect size was 0.079 (95% CI −0.09 to 0.25), 0.17 (−0.17 to 0.52) for passage comprehension, 0.01 (−0.25 to 0.28) for receptive language and −0.04 (95% −0.33 to 0.25) for expressive language. For comparisons with the treated control groups the equivalent pooled effect sizes were −0.026 (95% CI −0.40 to 0.35), −0.10 (−0.40 to 0.21) for passage comprehension, 0.02 (−0.27 to 0.31) for receptive language and −0.06 (−0.33 to 0.20) for expressive language. None of the eight pooled effect sizes were reliably different from zero, and four of the effect sizes were actually negative (indicating worse performance in the Fast ForWord treatment group than the control group). Thus, from the studies we have identified and analysed here there is no convincing evidence that Fast ForWord is effective in improving children's single word reading, passage reading comprehension, receptive language or expressive language skills.

no evidence from the analysis carried out that Fast ForWord is effective as a treatment for children's oral language or reading difficulties.

Tokpah (2008) 0.38

This meta-analysis sought to investigate the overall effectiveness of computer algebra systems (CAS) instruction in comparison to non-CAS instruction, on students' achievement in mathematics at pre-college and post-secondary institutions. The study utilised meta-analysis on 31 primary studies (102 effect sizes, N= 7,342) that were retrieved from online research databases and search engines, and explored the extent to which the overall effectiveness of CAS was moderated by various study characteristics. The overall effect size, 0.38, was significantly different from zero. The mean effect size suggested that a typical student at the 50th percentile of a group taught using non-CAS instruction could experience an increase in performance to the 65th percentile, if that student was taught using CAS instruction. The fail-safe N, Nfs, hinted that 11,749 additional studies with non-significant results would be needed to reverse the current finding. Three independent variables (design type, evaluation method and time) were found to significantly moderate the effect of CAS.

The current results do not predict future trends on the effectiveness of CAS; however, these findings suggest that CAS have the potential to

The average effect size for CAS in a tutorial role (d = 0.40) did not differ significantly from the average effect size for CAS in a tool role (d=0.39), ns.

The average effect sizes for studies that controlled for the effect of teacher (different teachers) and studies that did not control for the effect of teacher (same teacher) were found to be 0.41 and 0.30, respectively.

The average effect size for studies in which CAS were used during evaluation (d = 0.31) was significantly lower than the average effect size for studies in which CAS were not used during evaluation (d = 0.42), $QB(1) = 4.35$, $p<0.05$

The average effect size for studies conducted from 1990 to 1999 (d = 0.51) was significantly larger than the average effect size for studies conducted from 2000 to 2007 (d = 0.24), $\chi^2 (1) = 27.78$, $p< 0.05$.

Whilst no other pair of comparison was significant, the difference between the average effect size for studies conducted in the 1980's (d = 0.34) was less than that of studies conducted from 1990 to 1999.

Published studies (d = 0.38) unpublished studies (d = 0.39) ns

Appendix A.5.3: (cont.)

Title	Overall ES	Abstract	Moderator variables
		improve learning in the classroom. Regardless of how CAS were used, the current study found that they contributed to a significant increase in students' performance.	
Torgerson and Zhu (2003)	0.890 (CI 0.245 to 1.535) – word processing on writing 0.204 (CI –0.168 to 0.576) – ICT on spelling –0.047 (CI –0.33 to 0.236) – computer texts on reading comprehension/ questioning effect size 0.282 CI – 0.003 to 0.566) Computer texts on reading comprehension/ story retelling	What is the evidence for the effectiveness of ICT on literacy learning in English, 5–16? Studies were retrieved from the three electronic databases. PsycInfo and ERIC were the richest sources for retrieving RCTs for this review. 5.1.2 Mapping of all included studies Forty-two RCTs were identified for the effectiveness map. 5.1.3 Nature of studies selected for effectiveness in-depth review The 12 included RCTs were assessed as being of 'medium' or 'high' quality in terms of internal quality: 'high' quality in terms of relevance to the review; 'medium' or 'high' in terms of the relevance of the topic focus; and 'medium' or 'high' for overall weight of evidence. All 12 studies were undertaken in the US with children aged between 5 and 14. Seven of the RCTs included samples where all or half of the participants experienced learning disabilities or difficulties or specific learning disabilities. All 12 studies focused on the psychological aspects or representations of literacy.	A range of five different kinds of ICT interventions emerged from the 12 included RCTs in the review: (1) computer-assisted instruction (CAI), (2) networked computer system (classroom intranet), (3) word-processing software packages, (4) computer-mediated texts (electronic text) and (5) speech synthesis systems. There were also three literacy outcomes: (1) reading, including reading comprehension and phonological awareness (pre-reading understandings), (2) writing and (3) spelling. Six RCTs evaluated CAI interventions The CAI interventions consisted of studies designed to increase spelling abilities, reading abilities or phonological awareness (pre-reading understandings). One RCT evaluated a networked computer system intervention and two RCTs evaluated word-processing interventions; three RCTs evaluated computer-mediated texts interventions and one RCT evaluated a speech synthesis intervention. In synthesis (1), for five different

Study	Effect size	CI		
Torgerson and Elbourne (2002)	0.37	CI −0.02 to 0.77	Recent government policy in England and Wales on information and communication technology (ICT) in schools is heavily influenced by a series of non-randomised controlled studies. The evidence from these ICT interventions, overall we included 20 comparisons from the 12 RCTs: 13 were positive and seven were negative. Of the positive ones, three were statistically significant, whilst of the seven negative trials, one was statistically significant. These data would suggest that there is little evidence to support the widespread use of ICT in literacy learning in English. This also supports the findings from previous systematic reviews that have used data from rigorous study designs. It also supports the most recent observational data from the Impact2 study. These findings support the view that ICT use for literacy learning should be restricted to pupils participating in rigorous, randomised trials of such technology. In synthesis (2), we undertook three principal meta-analyses: one for each of the three literacy outcomes measures in which we were interested. In two, there was no evidence of benefit or harm; that is, in spelling and reading the small effect sizes were not statistically significant). In writing, there was weak evidence for a positive effect, but it was weak because only 42 children altogether were included in this meta-analysis.	No UK RCTs found. Our review found that the evidence base for the teaching of spelling by using a computer was very weak. It was particularly surprising that so few randomised controlled trials had been

Appendix A.5.3: (cont.)

Title	Overall ES	Abstract	Moderator variables
		evaluations is equivocal with respect to the effect of ICT on literacy. In order to ascertain whether there is any effect of ICT on one small area of literacy, spelling, a systematic review of all randomised controlled trials (RCTs) was undertaken. Relevant electronic databases (including BEI, ERIC, Web of Science, PsycINFO, The Cochrane Library) were searched. Seven relevant RCTs were identified and included in the review. When six of the seven studies were pooled in a meta-analysis there was an effect, not statistically significant, in favour of computer interventions (effect size = 0.37, 95% confidence interval = −0.02 to 0.77, p = 0.06). Sensitivity and subgroup analyses of the results did not materially alter findings. This review suggests that the teaching of spelling by using computer software may be as effective as conventional teaching of spelling, although the possibility of computer-taught spelling being inferior or superior cannot be confidently excluded due to the relatively small sample sizes of the identified studies. Ideally, large pragmatic randomised controlled trials need to be undertaken.	undertaken in this area. This lack of evidence of effectiveness should not be interpreted as evidence that computer spelling programmes should instantly be withdrawn – the quality of the trials was variable, there may be unmeasured benefits and there is no evidence that the programmes will harm children's spelling. Nevertheless, these conclusions are based on the best-available research appropriate to answering questions about the effectiveness of ICT on teaching and learning spelling. The onus should be on those wishing to introduce interventions such as these to first evaluate them formally, using rigorous research methods (large pragmatic RCTs), illuminated by the relevant theoretical developments.
Waxman et al. (2002)	0.39 on cognitive outcomes CI −0.050 to 0.830	To estimate the effects of teaching and learning with technology on students' cognitive, affective and behavioural outcomes of learning,	The relationship of each of the 56 conditioning (i.e., independent) variables to the mean study weighted effect size was tested for significance

138 effect sizes were calculated using statistical data from 20 studies that contained a combined sample of approximately 4,400 students. The mean of the study-weighted effect sizes averaging across all outcomes was .30 ($p<0.05$), with a 95% confidence interval (CI) of 0.004–0.598. This result indicates that teaching and learning with technology has a small, positive, significant ($p<0.05$) effect on student outcomes when compared to traditional instruction. The mean study-weighted effect size for the 13 comparisons containing cognitive outcomes was 0.39, and the mean study-weighted effect size for the 60 comparisons that focused on student affective outcomes was 0.208. On the other hand, the mean study-weighted effect size for the 30 comparisons that contained behavioural outcomes was –0.154, indicating that technology had a small, negative effect on students' behavioural outcomes. The overall study-weighted effects were constant across the categories of study characteristics, quality of study indicators, technology characteristics and instructional/teaching characteristics.

using ANOVA. The results indicate that none of the variables had a statistically significant ($p<0.01$) impact on the study-weighted effect size. In other words, the overall findings suggest that the results do not differ significantly across categories of technology, instructional characteristics, methodological rigour, characteristics of the study and subject characteristics

Cheung and Slavin (2012)

0.16 SE = 0.02 CI [0.12, 0.21]

The purpose of this review is to learn from rigorous evaluations of alternative technology applications how features of using technology programmes and characteristics of their evaluations affect reading outcomes for students in grades K–12. The review applies consistent

Type of intervention:
computer-managed learning = 0.19
innovative technology approach = 0.18
comprehensive = 0.28
supplemental = 0.11
Programme intensity:

Appendix A.5.3: (*cont.*)

Title	Overall ES	Abstract	Moderator variables
		inclusion standards to focus on studies that met high methodological standards. A total of 84 qualifying studies based on over 60,000 K–12 participants were included in the final analysis. Consistent with previous reviews of similar focus, the findings suggest that educational technology applications generally produced a positive, though small, effect (ES = +0.16) in comparison to traditional methods. There were differential impacts of various types of educational technology applications. In particular, the types of supplementary computer-assisted instruction of supplementary computer-assisted instruction programmes that have dominated the classroom use of educational technology in the past few decades were not found to produce educationally meaningful effects in reading for K–12 students (ES = +0.11), and the higher the methodological quality of the studies, the lower the effect size. In contrast, innovative technology applications and integrated literacy interventions with the support of extensive professional development showed more promising evidence. Although many more rigorous, especially randomised, studies of newer applications are needed, what unifies the methods found in this review to have great promise is the use of technologies in close connection with teachers' efforts.	high = 0.19, low = 0.11 Implementation level: low = 0.01, medium = 0.18, high = 0.22, NA = 0.16 Grade levels: kindergarten = 0.15 elementary = 0.10 secondary = 0.31 SES: low = 0.17, high = 0.12 Gender: M = 0.28, F = 0.12 Race: African American = 0.12 Hispanic = 0.42 White = 0.11 Year of publication: 1980s = 0.16 1990s = 0.08 2000s = 0.18 2010s = 0.17 Research design: randomised:0.08 randomised quasi = 0.16 matched = 0.19 matched post-hoc = 0.19 Sample size: large = 0.13, small = 0.25 Design and size:

| Wouters et al. (2013) | d = 0.29 (learning)
SE = 0.06
CI 0.17 to 0.42 | It is assumed that serious games influences learning in two ways, by changing cognitive processes and by affecting motivation. However, until now research has shown little evidence for these assumptions. We used meta-analytic techniques to investigate whether serious games are more effective in terms of learning and more motivating than conventional instruction methods (learning: $k = 77$, $N = 5{,}547$; motivation: $k = 31$, $N = 2{,}216$). Consistent with our hypotheses, serious games were found to be more effective in terms of learning ($d = 0.29$, $p < 0.01$) and retention ($d = 0.36$, $p < 0.01$), but they were not more motivating ($d = 0.26$, $p\_0.05$) than conventional instruction methods. Additional moderator analyses on the learning effects revealed that learners in serious games learned more, relative to those taught with conventional instruction methods, when the game was supplemented with other instruction methods, when multiple training sessions were involved and when players worked in groups. | large randomised = 0.07
small randomised = 0.21
large matched = 0.16
small matched = 0.24
Publication type:
published = 0.25
unpublished = 0.14
Type of instruction:
active = 0.28
passive = 0.06
mixed = 0.50
Computer game alone:
inclusive = 0.41
exclusive = 0.20
Sessions:
1 session = 0.10
multiple sessions = 0.54
Group:
individual = 0.22
group = 0.66
Domain:
biology = 0.11
math = 0.17
language = 0.66
engineering = −0.36
others = 0.54
Age:
children = 0.30
preparatory education = 0.33
students = 0.23 |

Appendix A.5.3: *(cont.)*

Title	Overall ES	Abstract	Moderator variables
			adults = 0.50 Narrative: yes = 0.25 no = 0.46 Publication source: peer review = 0.36 proceedings = −0.16 unpublished = −0.20 Randomisation: yes = 0.08, no = 0.44 Design: post-test only = 0.25 pre-post = 0.32
Blok et al. (2002)	0.19 SE = 0.05	How effective are computer-assisted instruction (CAI) programmes in supporting beginning readers? This article reviews 42 studies published from 1990 onward, comprising a total of 75 experimental comparisons. The corrected overall effect size estimate was d = 0.19 (± 0.06). Effect sizes were found to depend on two study characteristics: the effect size at the time of pre-testing and the language of instruction (English or other). These two variables accounted for 61% of the variability in effect sizes. Although an effect size of d = 0.20 shows little promise, caution is needed because of the poor quality of many studies.	We found two study characteristics to be related to study outcomes. Effect sizes were higher when (a) the experimental group displayed an advantage at the pre-test, and (b) the language of instruction was English. The effects of these two predictors reduced the variability of the study outcomes by a sizable 61%. Several other study characteristics appeared not to be related to study outcomes. Among these were design characteristics (subject assignment, size of experimental group, type of post-test score), population characteristics (regular or dyslexic readers, mean age of students) and treatment characteristics (type of

Study	Comparison	Description	Results
Clark et al. (2015)	*digital games vs. non-game conditions g = 0.33 CI 0.19 to 0.48 *value-added comparisons g = 0.34 CI 0.17 to 0.51	In this meta-analysis, we systematically reviewed research on digital games and learning for K–16 students. We synthesised comparisons of game versus non-game conditions (i.e., media comparisons) and comparisons of augmented games versus standard game designs (i.e., value-added comparisons). We used random-effects meta-regression models with robust variance estimates to summarise overall effects and explore potential moderator effects. Results from media comparisons indicated that digital games significantly enhanced student learning relative to non-game conditions (g = 0.33, 95% confidence interval [0.19, 0.48], k = 57, n = 209). Results from value-added comparisons indicated significant learning benefits associated with augmented game designs (g = 0.34, 95% confidence interval [0.17, 0.51], k = 20, n = 40). Moderator analyses demonstrated that effects varied across various game mechanics characteristics, visual and narrative characteristics, and research quality characteristics. Taken together, the results highlight the affordances of games for learning as well as the key role of design beyond medium.	experimental programme, programme length, programme duration). Digital vs. nongame: cognitive learning = 0.35 Number of game sessions: single = 0.08, multiple = 0.44 Game included nongame instruction: yes = 0.36, no = 0.32 Game players: single = 0.45, single competitive = −0.06, collaborative team = 0.22 Game type: adding points = 0.53, more than points = 0.25 Variety of game actions: small = 0.35, medium = 0.43, large = 0.40 Intrinsic/extrinsic type: not fully intrinsic = 0.33, intrinsic = 0.19, simplistically intrinsic = 0.49 Research design: experimental = 0.28, quasi = 0.50 Further moderators include technical variables
Kunkel (2015)	A. CAI vs. no-treatment control (post-test)	The purpose of this study was to investigate the effectiveness of computer-assisted instruction (CAI) to improve the reading outcomes of students in preschool through high school. A	School grade: Dataset A: preschool/kindergarten +0.50 elementary (K–6) +0.23 Secondary (6–12) +0.03

Appendix A.5.3: (*cont.*)

Title	Overall ES	Abstract	Moderator variables
	g = 0.21, SE = 0.06, CI 0.09 to 0.33 B. CAI vs. no-treatment (gain) g = 0.13, SE = 0.02 CI 0.10 to 0.16 C. CAI vs. teacher led (post) g = −0.05, SE = 0.06 CI −0.16 to 0.07	total of 61 studies met criteria for this review, and 101 independent effect sizes were extracted. Results indicated that the mean effects for students receiving reading CAI were small, positive and statistically significant when compared to control groups receiving no treatment or non-reading CAI. Categorical moderator analyses and meta-regression were conducted to explore the variation in effects. Results of an analysis of research quality indicated that, on average, about half of quality indicators were met. The results of this meta-analysis show that CAI in reading can effectively enhance the reading outcomes of students in preschool through high school. Future, high-quality research should be conducted to identify effective programmes and establish best practice in the instructional design of CAI to enhance the reading skills of all students.	Dataset B preschool/kindergarten +0.20 elementary (K–6) +0.12 secondary (6–12) +0.18 Dataset C preschool/kindergarten +0.04 elementary (K–6) +0.21 secondary (6–12) +0.02 Reading outcomes: Dataset A: phonemic awareness +0.33 phonics +0.24 fluency +0.17 vocabulary −0.13 comprehension +0.20 broad reading +0.28 Dataset B phonemic awareness+0.33 phonics +0.11 fluency -0.20 vocabulary +0.05 comprehension +0.13 broad reading +0.14 Dataset C phonemic awareness−0.07 phonics −0.15 0 fluency +0.24 vocabulary −0.21 comprehension +0.27 broad reading +0.48

| Zheng et al. (2016) | Reading d = 0.12 SE = 0.06 CI −0.08 to 0.24 English language arts d = 0.15, SE = 0.06 CI 0.02 to 0.27 Writing d = 0.20, SE0.09 0.00 to 0.39 Science d = 0.25, SE = 0.11 CI 0.02 to 0.47 Math d = 0.16 SE = 0.08 CI 0.00 to 0.32 | Over the past decade, the number of one-to-one laptop programmes in schools has steadily increased. Despite the growth of such programmes, there is little consensus about whether they contribute to improved educational outcomes. This article reviews 65 journal articles and 31 doctoral dissertations published from January 2001 to May 2015 to examine the effect of one-to-one laptop programmes on teaching and learning in K–12 schools. A meta-analysis of 10 studies examines the impact of laptop programmes on students' academic achievement, finding significantly positive average effect sizes in English, writing, mathematics and science. In addition, the article summarises the impact of laptop programmes on more general teaching and learning processes and perceptions as reported in these studies, again noting generally positive findings. | meta-analyses on 10 studies of the 96. ages K–12 *no moderators *no publication bias |

Appendix A.5.4: *Learning styles meta-analyses*

Title	Overall ES	Abstract	Moderator variables
Kavale and Forness (1987)	0.14 SE 0.06	A literature search identified 39 studies assessing modality preferences and modality teaching. The studies, involving 3,087 disabled and nondisabled elementary/secondary level subjects, were quantitatively synthesised. Subjects receiving differential instruction based on modality preferences exhibited only modest gains.	Researchers found that modality matched to instruction produced an overall effect size of 0.128 on achievement test performance.
Garlinger and Frank (1986)	−0.03	The effects on academic achievement associated with matching students and teachers on field-dependent-independent dimensions of cognitive style are investigated. Since research in this area is still in a developmental stage and the results are often conflicting, a meta-analysis was performed on the data provided by the studies reviewed in an attempt to integrate and clarify the current status of findings relevant to this issue. The meta-analysis technique used was the effect size method. A summary table of study characteristics and effect sizes is provided; implications of the results of the meta-analysis are discussed.	No clear trend is evident for the effect sizes produced for the studies reviewed, as they range from +0.47 to −0.99 for matching field-dependent to -1.42 for matching field-independent students, from +0.54 to −0.25 for overall matching effects.
Lovelace (2002)	0.67	The author performed a quantitative synthesis of experimental research conducted between 1980 and 2000, in which the Dunn and Dunn Learning-Style Model was used. Of the 695 citations elicited by the database and reference section searches, 76 original	Three indicators rejected homogeneity for achievement and attitude effect sizes. Mean effect sizes for achievement and attitude were variable enough to be described as heterogeneous. Therefore,

research investigations met the established inclusion criteria.

The 7,196 participants provided 168 individual effect sizes. The mean effect-size values calculated and interpreted through this meta-analysis provided evidence for increased achievement and improved attitudes when responsive instruction was available for diagnosed learning-style preferences.

Mean effect-size results for achievement from the present and previous meta-analyses were consistent. The author suggested that, on average, learning-styles responsive instruction increased the achievement or improved the attitudes towards learning, or both, of all students. Although several moderating variables influenced the outcome, results overwhelmingly supported the position that matching students' learning-style preferences with complementary instruction improved academic achievement and student attitudes towards learning. The Dunn and Dunn model had a robust moderate to large effect that was practically and educationally significant.

the author searched for variables that moderated the effect sizes; six were found. Five variables that had a moderating effect on achievement effect sizes were (a) publication type, (b) degree of preference, (c) school type, (d) academic level and (e) demographic region. Mean effect size was greater among published than unpublished studies.

Slemmer (2002)

To identify forms of technology or types of technology-enhanced learning environments that may effectively accommodate the learning needs of students, 48 studies were included in a meta-analysis to determine the effects of learning styles on student achievement within technology-enhanced learning environments. A total of 51 weighted effect sizes were calculated from these studies with moderator variables coded for five study characteristics, six methodology

0.13
SE 0.03
CI 0.08
to 0.18

Sample size: 1,040 = 0.14, 41–80 = 0.29, 80 and more = 0.07.
Type of publication: journal article = 0.19, ERIC doc = -0.02, dissertation/thesis = 0.07, other = 0.08.
Year of publication: 90–92 = 0.12, 93–95 = 0.20, 69–98 = 0.11, 99–01 = 0.06.

Appendix A.5.4: (*cont.*)

Title	Overall ES	Abstract	Moderator variables
		characteristics and six programme characteristics. This meta-analysis found that learning styles do appear to influence student achievement in various technology-enhanced learning environments, but not at an overall level of practical significance. The total mean weighted effect size for the meta-analysis was z_r = 0.1341. Although the total mean weighted effect size did not reach the established level of practical significance (z_r = 0.16), the value was greater than z_r = 0.10, which is the level generally established by researchers as having a small effect. Additional findings from the moderator variables included: (1) Articles published in journals were the only type of publication that produced a significant mean weighted effect size (z_r = 0.1939). (2) Studies that reported t statistics produced one of the highest total mean weighted effect sizes (z_r = 0.4936) of any of the moderator variables. (3) Studies that reported an F statistic with df = 1 in the numerator had a significant total mean weighted effect size (z_r = 0.2125); whilst studies that reported an F statistic with df > 1 in the numerator had a non-significant total mean weighted effect size (z_r = 0.0637). (4) When all of the students received the same technology-enhanced lesson, there was a significant difference in student achievement between students with different learning styles (z_r = 0.2952). (5) Studies that used Witkin's learning styles measure indicated a	Pre-test equivalence: unspecified = 0.29, statistical control = 0.21, random assignment = 0.04, statistical control and random assignment = 0.21, pre-test scores = 0.21, pre-test and random assignment = 0.39. Statistical power: unspecified = 0.13, probability threat = 0.16, adequately minimises = 0.12. Type of statistic: F statistic = 0.09, T statistic = 0.49. Type of treatment: same for all = 0.29, same with variations = 0.07, different technology methods = 0.21, technology vs. lecture = 0.09, distance education vs. live = 0.15. Type of delivery system: hypertext = 0.13, hypertext/multimedia = 0.06, web-based = 0.03, interactive Type of application: computer-assisted instruction = 0.09, computer simulation = 0.11, videotape/videodisc = 0.21, distance education = 0.34. Type of instruction for treatment: unspecified = 0.16, individual = 0.10, small group = 0.07, large group = 0.33, mixed = 0.10.

		significant interaction between students' learning style and technology-enhanced learning environments as measured by student achievement (z_r = 0.1873), whilst none of the quadrant-based learning style models indicated a significant interaction. (6) As the duration of treatment increased, the findings of the studies increased in significance. In general, this study provided evidence that under some conditions, students interact differently with technology in technology-enhanced learning environments depending on their specific learning style and the type of technology encountered.	Treatment duration: unspecified = 0.18, less than 1 week = 0.09, 1–4 weeks = −0.06, 1–4 months = 0.13, longer than 4 months = 0.71.
Tamir (1985)	0.06	Most of the articles and dissertations dealing with cognitive preferences which were written since the invention of the construct in the early 1960s have been reviewed. Fifty-four of them were found suitable for meta-analysis. The meta-analysis presents means and standard deviations of reliabilities, correlations, standard scores and effect sizes. The effects and relationships of cognitive preferences and important school and learning related variables were studied. The results provide baseline data for comparative purposes and offer empirical evidence which support the construct validity of cognitive preferences.	Cognitive preferences 54 studies 18 achievement SMD = 0.06 Mean correlation 0.0275 d = 0.06. The results indicate that, on the average, achievement is positively correlated with principles (r = 0.16) and questioning (r = 0.12), negatively with recall (r = −0.14) and practically not correlated with application (r = −0.03).
Kanadli (2016)	1.03	The purpose of this study is to calculate the effect size, by running a meta-analysis, of the experimental studies carried out in Turkey between 2004 and 2014 that investigate the effect of learning styles on academic achievement, attitude and retention and to define whether the academic achievement shows a significant	Under the random-effect model, the common effect sizes were as follows: studies employing the 4MAT System, 1.168 (0.860, 1.477); studies employing Perceptual Learning Style Model, 0.870 (0.653, 1.023); studies employing the

Appendix A.5.4: (cont.)

Title	Overall ES	Abstract	Moderator variables
		difference in terms of learning styles model, experimental design and course type. For this purpose, a meta-analytical review method was employed to combine the outcome of the independent experimental studies.	Dunn and Dunn Learning Style Model, 1.331 (1.047, 1.087); studies employing the Kolb Learning Styles Model, 1.067 (−0.876, 3.009). The model with the largest effect on academic achievement is the Dunn and Dunn Learning Style Model and the model with the smallest effect on academic achievement is the Perceptual Learning Style Model.
		The studies included in this review were collected from CoHE National Thesis Archive (2015), ULAKBIM (2015), Google Academic (2015), ERIC (2015) and EBSCO (2015) databases. As a result of the searching process, 402 studies were assessed according to the inclusion criteria and 30 experimental studies were included in this study. Cohen's d coefficient was calculated for the effect size in this study. Because there was a high amount of heterogeneity ($Q > x2$, $p<0.05$) among the effect sizes of the studies, the common effect size was calculated according to the random-effect model. As a result of meta-analysis, it was determined that the instructional designs based on the learning styles model had a large effect on the academic achievement (d = 1.029), attitude (d = 1.113) and retention (d = 1.290). Moreover, the academic achievement did not show any significant difference according to learning style model, course type and experimental design.	Studies conducted on the English course, 0.743 (0.332, 1.154); studies conducted on the natural science course, 1.165 (−0.836, 1.494); studies conducted on social science courses, 1.111 (0.664, 1.558); and studies conducted in informatics, 0.702 (0.060, 1.343). Studies with quasi-experimental design, 1.099 (0.890, 1.307); the studies with true experimental design, 0.647 (0.226, 1.068) The effect of the learning style models on academic achievement (Q = 0.016, p = 0.901) did not show a significant difference (p>0.05) according to the publication status.

Appendix A.6.1: *Reading comprehension meta-analyses*

Citation	Overall ES	Abstract	Moderator variables
Berkeley et al. (2010)	0.65 all measures Criterion referenced measures 0.69 CI 0.56 to 0.83 Norm referenced measures 0.52 CI 0.33 to 0.70	Meta-analysis procedures were employed to synthesise findings of research for improving reading comprehension of students with learning disabilities published in the decade following previous meta-analytic investigations. Forty studies, published between 1995 and 2006, were identified and coded. Nearly 2,000 students served as participants. Interventions were classified as fundamental reading skills instruction, text enhancements, and questioning – including those that incorporated peer-mediated instruction and self-regulation strategy instruction. Mean weighted effect sizes were obtained for criterion referenced measures: 0.69 for treatment effects, 0.69 for maintenance effects, and 0.75 for generalisation effects. For norm referenced tests, the mean effect size was 0.52 for treatment effects. These outcomes were somewhat lower than but generally consistent with those of previous meta-analyses in their conclusion that reading comprehension interventions have generally been very effective. Higher outcomes were noted for interventions that were implemented by researchers. Implications for practice and further research are discussed.	Grade level (criterion-referenced measures) Middle and high school students 0.80 Elementary school students 0.52 Setting (norm-referenced measures) One to one 0.53 CI −0.15 to 1.21 Small group 0.58 CI 0.17 to 0.78 Passage type (norm-referenced measures) Narrative 0.88 CI 0.36 to 1.40 Expository 0.32 CI −0.02 to 0.65 Both 0.72 CI −2.45 to 3.50 Unknown(0.28 CI −2.27 to 2.82 Treatment duration (norm-referenced measures) Medium 0.49 CI 0.14 to 0.84 Long 0.53 CI 0.27 to 0.79 Study design (norm-referenced measures) Experimental 0.38 CI −0.10 to 0.86 Quasi-exp. 0.45 CI 0.18 to 0.72 Pre-post 0.98 CI −1.74 to 3.69 Treatment Delivery: Researchers 0.83 CI 0.62 to 1.04 Teachers 0.56 CI 0.33 to 0.79 Adult Tutors 0.42 Technology 0.66 CI −2.90 to 4.72 40 studies N = 1734

Appendix A.6.1: (cont.)

Citation	Overall ES	Abstract	Moderator variables
Davis (2010)	0.36 CI 0.21 to 0.51 SE 0.08	This meta-analytic review includes intervention studies published between 1980 and 2009 in which students in grades 4–8 are taught to use two or more comprehension strategies. The collected studies were coded using a systematic data extraction scheme developed to address the central questions of the review. Information related to the characteristics of the student sample and instructional and methodological characteristics of each study were compiled in a database. Numerical effect sizes for each study for each major outcome measure were computed. The mean effect of comprehension strategy instruction on each of the targeted outcome constructs was calculated to provide an overall summary of instructional effectiveness.	A few of the other name-brand frameworks appear to be equally or more effective than RT at improving middle grades comprehension achievement. These include peer-assisted learning strategies, Think-aloud instruction, transactional strategies instruction and concept-oriented reading Instruction. However, these effects are based on limited available evidence. Overall impact on standardised tests of reading comprehension, but significant heterogeneity (I-squared 53.57). Reciprocal teaching 0.31
Edmonds et al. (2009)	0.47 CI 0.12 to 0.82	This article reports a synthesis of intervention studies conducted between 1994 and 2004 with older students (grades 6–12) with reading difficulties. Interventions addressing decoding, fluency, vocabulary, and comprehension were included if they measured the effects on reading comprehension. Twenty-nine studies were located and synthesised. Thirteen studies met criteria for a meta-analysis, yielding an effect size (ES) of 0.89 for the weighted average of the difference in comprehension outcomes between	Measurement type All measures 0.89 CI 0.42 to 1.36 Standardised measurement 0.47 CI 0.12 to 0.82 Researcher developed measures 1.19 CI 1.10 to 1.37 Intervention Type Fluency –0.03 CI –0.56 to 0.62 Word study 0.34 CI –0.22 to 0.88 Multicomponent 0.72 CI 0.45 to 0.99 Comprehension 1.23 CI 0.96 to 1.50

	treatment and comparison students. Word-level interventions were associated with ES = 0.34 in comprehension outcomes between treatment and comparison	
Elleman et al. (2009)	Standardised measures 0.10	A meta-analysis of vocabulary interventions in grades pre-K to 12 was conducted with 37 studies to better understand the impact of vocabulary on comprehension. Vocabulary instruction was found to be effective at increasing students' ability to comprehend text with custom measures (d = 0.50), but was less effective for standardised measures (d = 0.10). When considering only custom measures, and controlling for method variables, students with reading difficulties (d = 1.23) benefited more than three times as much as students without reading problems (d = 0.39) on comprehension measures. Gains on vocabulary measures, however, were comparable across reading ability. In addition, the correlation of vocabulary and comprehension effects from studies reporting both outcomes was modest (r = 0.43).	Differences were found in the pattern of effects for vocabulary and comprehension. The overall comprehension effect for students with reading difficulties (d = 1.23) was much larger than that for students with no indicated problem (d = 0.39). However, for vocabulary outcomes, both groups made similar gains from instruction (d = 0.84 for students with no indicated problem and d = 0.79 for students with reading difficulties). The corrected correlation of 0.74 means that roughly 55% of the variance in comprehension gains may be explained by vocabulary growth.
Fauzan (2003)	0.50 CI 0.45 to 0.56 0.55 CI 0.48 to 0.63 (with outliers excluded)	The purpose of the study was to investigate the effectiveness of metacognitive strategies on reading comprehension by means of (a) a meta-analysis and (b) an experiment designed following the meta-analysis implemented in Sarawak, Malaysia. Before the meta-analysis, the prevalent theories and issues in the reading literature such as metacognition, models of reading,	Features considered were; publication, grade and ability levels and design characteristics such as methods of assignment, length of treatment and the instructional variables. The findings suggests that the use of metacognitive strategies with other instructional feature such as motivation improved the learners' reading comprehension by 0.76, depending on the students' grade and

Appendix A.6.1: (cont.)

Citation	Overall ES	Abstract	Moderator variables
		measurements, motivation and previous meta-analysis were discussed to provide a better understanding of the research area in this study. A meta-analytic procedure conducted to review the primary research studies of metacognitive strategies used effect size as the measure of effectiveness. Searching for the articles and theses in the 1980s until 2001 yielded a record of 473 abstracts and articles from which there were 27 studies with a total number of 82 effect sizes that could be quantitatively synthesised to compare the group performance of the experimental and control groups. The weighted effect size was 0.50 (CI 0.45 to 0.56) when dependent effect sizes were synthesised, and 0.55 (CI 0.48 to 0.63) when the extreme 'outliers' or deviated effect sizes were excluded and independent effect sizes were created. Overall, the effect size was moderate indicating a positive outcome of the metacognitive strategies. The effect sizes were not homogeneous and further analyses of the qualitative and quantitative features of the studies were made to develop possible reliable estimates.	ability level. The effect was 0.70 when studies was conducted with the mixed ability group. However, the effect sizes were equally effective for younger students at grade 2–3 and those at the college where the effect sizes were 0.69 and 0.72. The smallest effect was on grade 10–11 students with an effect size of 0.23. Another important feature that might influence reading comprehension is the type of reading materials used. The effect was 0.70 when expository text was used compared to 0.50 with narrative text. The issues surrounding the instructional time remain unresolved for the elapsed time of <15 days, 15–25 days and >25 days were equally moderate at 0.49, 0.47 and 0.61 and the number of sessions of 1–5, 6–10 and >10 at 0.63, 0.65 and 0.40. Considering the students' ability level, perhaps extended guided practice, distributed within the time given for the research would improve the reading strategies used and subsequently improves learners' reading comprehension
Fukkink and Glopper (1998)	0.43	A meta-analysis of 21 instructional treatments aimed at enhancing the skill of deliberately deriving word meaning from context during reading shows a medium effect size of 0.43 standard deviation units (p<0.000). An	

exploratory multilevel regression analysis shows that clue instruction appears to be more effective than other instruction types or just practice (β = 0.40). Effect size correlates negatively with class size (β = .03). Implications for instruction and future research are discussed. Future studies should investigate the effect of instruction on both the skill of deriving word meaning from context and incidental word learning to evaluate its contribution to vocabulary growth.

Galloway (2003)   0.74

Over the past 30 years, research has increasingly sought to examine the efficacy of metacognitive strategy instruction to improve reading comprehension. Whilst some interventions have focused on single-strategy interventions, others have employed multiple-component strategy packages to improve the self-regulatory skills of readers. Reciprocal teaching is the most widely researched multi-component metacognitive strategy-training program. Although an early review of the reciprocal teaching procedure was conducted in 1994 by Rosenshine & Meister, it was based primarily on unpublished work. Since that time, the number of published studies examining the reciprocal teaching procedure has more than doubled. In addition, recent advances in the evaluation of the evidence base for interventions in school psychology have helped to delineate the variables important for reviewing interventions in education and psychology.

Using a traditional meta-analysis in conjunction with recently developed standards for evaluating

Design
Between d = 0.77
Small-N d = 0.63
Grade:
Late elementary d = 0.59,
Early secondary d = 1.0
Post-secondary d = 0.59
Reader Skill:
Above average = 0.20
Average = −0.28,
Below average d = 0.39
Learning difficulties d = 0.76
EAL d = 1.10
Heterogeneous d = 0.79

Appendix A.6.1: (*cont.*)

Citation	Overall ES	Abstract	Moderator variables
		evidence-based interventions in school psychology, this study found a moderate effect size for interventions employing the reciprocal teaching procedure to improve reading comprehension. Unlike the earlier review of the reciprocal teaching procedure, this study did not find significant differences between the effect sizes produced for norm-referenced and experimenter/teacher-generated tests. Analysis of measures of strategy use and reading comprehension follow-up measures suggest that the effects of reciprocal teaching are not only a function of strategy use but are maintained over time. Whilst the certainty with which conclusions can be drawn from this study is limited due to a relatively small sample size, the reciprocal teaching procedure appears to hold promise for helping students to develop the types of self-regulatory strategies used by skilled readers to promote reading comprehension. Whilst there remain a number of questions regarding the conditions under which reciprocal teaching is maximally effective, the available evidence suggests that the procedure can help readers to develop skills that promote independent reading comprehension. Future research that investigates permutations of the procedure, its utility with populations with varying demographic characteristics, and the relationship between this procedure and other forms of reading instruction is likely to promote greater understanding of the procedure and its effects.	

| Scammacca et al. (2015) | 0.49 CI 0.38 to 0.60 on reading
0.21 SE 0.06 on reading comprehension | This meta-analysis synthesises the literature on interventions for struggling readers in grades 4 through 12 published between 1980 and 2011. It updates Scammacca et al.'s analysis of studies published between 1980 and 2004. The combined corpus of 82 study-wise effect sizes was meta-analysed to determine (a) the overall effectiveness of reading interventions studied over the past 30 years, (b) how the magnitude of the effect varies based on student, intervention and research design characteristics, and (c) what differences in effectiveness exist between more recent interventions and older ones. The analysis yielded a mean effect of 0.49, considerably smaller than the 0.95 mean effect reported in 2007. The mean effect for standardised measures was 0.21, also much smaller than the 0.42 mean effect reported in 2007. The mean effects for reading comprehension measures were similarly diminished. Results indicated that the mean effects for the 1980–2004 and 2005–2011 groups of studies were different to a statistically significant degree. The decline in effect sizes over time is attributed at least in part to increased use of standardised measures, more rigorous and complex research designs, differences in participant characteristics, and improvements in the school's 'business-as-usual' instruction that often serves as the comparison condition in intervention studies. | Type of intervention:
Reading comprehension 0.21
Fluency 0.28
Multiple component 0.11
Grade grouping
4th–5th 0.22
6th–8th 0.16
9th–12th 0.09
Type of implementer:
Researchers 0.14
Teacher 0.13
LD status:
All LD = 0.15
Some LD some struggling 0.12
Hours of intervention:
6–15 hours 0.22
16–25 hours 0.21
26+ hours 0.10
Design type:
Multiple treatment 0.07
Treatment versus control 0.17 |

Appendix B Types of Effect Size

In this book, I have concentrated on the use of a standardised mean difference, as this is most frequently used in intervention research to compare outcomes between two groups, across different measures. However, the description 'effect size' is used to describe a wide range of statistical approaches to compare groups and relationships between groups. The term can refer to a difference which is 'standardised' or 'unstandardised'. A simple or unstandardised difference is simply the difference between two means on the same test. This is fine for a simple comparison or if we always use the same test. The purpose of standardising is to relate the specific difference between groups to the distribution, so as to make the differences more comparable across different measures. It effectively removes the units the scores were measured in, such as a particular test, and replaces these with the spread of the scores. This is important for meta-analysis when we need to combine results from different tests and different studies. A standardised normal distribution is associated with standard probabilities so enables a number of further inferences to be drawn about the benefit across a group. Even here, however, there are different ways to calculate a standardised mean difference:

- Glass's Δ: where the effect size is standardised by the standard deviation of the control group;
- Cohen's d: where the effect size is standardised by the standard deviation of both groups;
- Hedges' g: where the effect size is standardised by the pooled standard deviation of both groups, but with an added factor to control against bias for very small samples (under 20).

Hedges' g can also be calculated from an analysis of variance (ANOVA) using the square root of the mean square error, when testing for differences between the two groups. It can be calculated from other statistics such as a t-test or a correlation (r). Similar procedures can be used to calculate equivalent effect size estimates from other analyses, such as

ANCOVA and multi-level models (Borenstein et al., 2008). You can calculate a measure of uncertainty around an effect size and use this to create confidence intervals, which relate to a conventional test of statistical significance.

Other kinds of effect sizes can express the relationships between two groups as correlations and the extent of overlap or degree of dependence (such as Pearson's r or the product-moment correlational coefficient which is a measure of the linear correlation between two variables). In medicine, it is more common to see effect sizes based on binary measures, such as the odds ratio (the odds of success in the treatment group relative to the odds of success in the control group) or risk ratio (the ratio of the probability of an event occurring (such as developing a disease, compared with to the probability of the event occurring in a comparison group).

Each of these different kinds of effect size can be converted into another, either directly (such as from a correlation to a standardised mean difference) or using assumptions about based on the distributions (such as an odds ratio to a standardised mean difference). However, this does not mean that they are directly comparable. The effect of a homework intervention, calculated from the difference in a randomised controlled trial between two groups, is of a different kind from the correlation between the amount of homework pupils do and how well they do at school. They both express a relationship between homework and learning, but one aims to identify a causal effect of deliberately introducing homework; the other may be due to other factors, such as that more effective schools set more homework, or that higher performing schools have the kind of pupils (and parents) who are willing to undertake homework.

In a meta-analysis, you need to be clear about what kind of relationship the effect size measures. This may be correlational, or there may be good causal grounds for seeing the difference as an effect. I don't like calling correlational relationships 'effects'. They express the effect of a variable in a mathematical equation, but this may or may not be an effect in terms of intentional change. Adding them together may make sense if you want to know what the overall average of the overall relationship between homework and learning is, but I prefer to keep causal inferences separate from relationships where there may be no causal association.

Appendix C Interpreting a 'Forest Plot'

This kind of visual representation succinctly combines much of the information from a meta-analysis in graphical form. Usually studies are listed in the left-hand column (either in alphabetical or date order). The central column (in the examples below or on the right-hand side in most software outputs) plots of the effect (usually a standardised mean difference in education research) for each of these studies (usually represented by a square) with confidence intervals presented as whiskers. The area of each square is proportional to the study's *weight* in the meta-analysis. The overall pooled effect is usually represented on the plot as a dashed vertical line (relative to a line of no effect) and as diamond at the bottom of the central column in the example below, with the lateral points of the diamond indicating the confidence intervals for the pooled estimate.

In the example below (Figure C1: see Chapter 1), the studies have also been divided into two subgroups to make a comparison between the impact of phonics interventions on normally developing children ('ability=1' at the bottom of the diagram) and the impact of pupils performing less well at reading than expected ('ability=0') in the top half. The subtotals can then easily be compared with the overall average using the dotted line. The research team used the software STATA to undertake this part of their analysis and to display the results.

Some of the information is repeated in the diagram, such as the weighting for each study, evident in the size of the square and '% Weight' column, the confidence intervals (indicated in the '95% CI' column) and either the 'whiskers' for each black square or the points of each diamond for the subtotals and overall average.

In the annotated example below (Figure C2) the forest plot has been labelled for comparison.

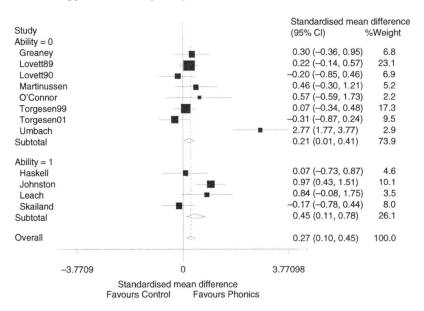

Figure C1: Forest plot from Torgerson et al. (2006)

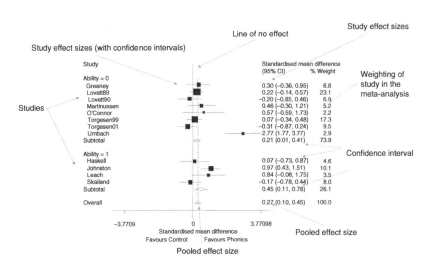

Figure C2: Annotated version of the forest plot from Torgerson et al. (2006)

Glossary

Bias:	A term denoting that a known or unknown variable is, or may be, responsible for an observed effect: see *publication bias*.
Cluster randomised trial:	In trials using this form of design the unit of allocation is at group level, e.g., class, school.
Co-variates or confounders:	These are variables that are associated with outcome (see *meta-regression*).
Coding:	Procedure that consists in identifying and extracting from primary studies information necessary to perform a meta-analysis.
Confidence intervals:	These indicate the level of uncertainty surrounding an effect size. The point estimate of effect of any intervention will always be imprecise. The level of the imprecision is dependent upon the sample size and event rate in the treatment groups. The use of confidence intervals (usually 95%, but sometimes 99% or 90%) reflects this imprecision in the study results. The assumptions include random sampling from a population, which is rarely achieved in education research.
CONSORT:	Consolidated Standards for Reporting Trials is the methodological standard adopted by many medical journals for publication of randomised controlled trials.
Controlled trial (CT):	This usually means a study with a control group that has been formed by means other than randomisation. Consequently, the validity of the study using this design is potentially threatened by selection bias.
Cumulative meta-analysis:	It is a meta-analysis that is performed first with one study, then with two studies, then with three

	studies, and so on, until all studies have been included.
Effect size:	This is a measure of the magnitude of a relationship or a difference between groups. It can be based on means (raw unstandardised mean difference or standardised mean difference (Cohen's d or Hedges' g, etc.) as well as other measures such as binary data (risk ratio, odds ratio, risk difference, etc.) and correlations (r); survival data (hazard ratio).
Exclusion criteria:	Criteria that define which studies should be excluded from a meta-analysis.
Fixed-effect model:	Statistical model used to combine the study effect sizes. According to this model (which is opposed to the random-effects model) there is a true effect size common to all the studies. In assigning a weight to each study, it takes into account only one source of variance: the within-study variance.
Forest plot:	This succinctly combines the information from a meta-analysis in graphical form. Usually studies are listed in the left-hand column (either in alphabetical or date order). The right-hand column plots of the effect (usually a standardised mean difference in education research) for each of these studies (often represented by a square) with confidence intervals represented as whiskers. The area of each square is proportional to the study's *weight* in the meta-analysis. The overall pooled effect is usually represented on the plot as a dashed vertical line and as diamond at the bottom of the right-hand columns with, the lateral points indicating the confidence intervals for the pooled estimate.
Funnel plot:	A method of assessing whether there is any probable publication bias. The effect size of each study is plotted against its sample size. Small studies will tend to have larger random variations in their effect sizes, which will be scattered along the x-axis close to the bottom of the y-axis. Larger studies will be higher up on the y-axis

and less scattered along the *x*-axis. A review with no publication bias would be expected to show a plot in the shape of an inverted funnel.

Grey literature: Research, such as technical reports or working papers, that is produced in print and electronic formats by a range of oranisations but which is not controlled and maintained by commercial publishers such as academic journals and publishing houses. It often has limited dissemination and, because it is hard to find, is often missing from meta-analyses.

Heterogeneity: This is a term used to refer to variation in effects between studies included in a meta-analysis. A certain amount of variation can be expected, according to sampling theory. This can be assessed in a meta-analysis and explored to see whether there is greater than expected variation and whether this can be explained by features of the included studies which have been coded (such as methodological heterogeneity from aspects of the design or pedagogical heterogeneity from differing characteristics of the intervention).

Inclusion criteria: Criteria that define which studies should be included in a meta-analysis.

Meta-analysis: A meta-analysis is a method of statistically combining the quantitative results of two or more studies. It aims to identify an overall average or 'pooled effect'. There are different ways to do this (see fixed effect and random effects models). It will usually include an exploration what explains variation in effects (see heterogeneity)

Fixed-effect model: The fixed-effect model of meta-analysis assumes that the variability is exclusively because of random sampling variation around a fixed effect. This is similar to weighting by sample size but uses the inverse variance model (each study is weighted by the reciprocal of it standard error).

Meta-regression: Statistical analysis used to test the effect of a continuous moderator.

Moderator: Variable that might explain differences in the effect sizes. If the moderator is categorical, its

	effect is tested by a subgroup analysis; if the moderator is continuous, its effect is tested by a meta-regression.
PRISMA:	Preferred Reporting Items for Systematic Reviews and Meta-analyses (PRISMA); an evidence-based minimum set of items for reporting in systematic reviews and meta-analyses. PRISMA focuses on the reporting of reviews evaluating randomised trials but can also be used as a basis for reporting systematic reviews of other types of research, particularly evaluations of interventions.
Publication bias:	This occurs when published studies differ systematically from unpublished studies (grey literature). Different methods for evaluating the publication bias are available (i.e., funnel plot, Rosenthal's fail-safe N, Orwin's fail-safe N, Duval and Tweedie's trim and fill method, etc.). Research which produces negative or null effects are less likely to be published than positive trials. Unless a systematic review includes these negative trials it can give a misleadingly optimistic assessment of the intervention. Because of reliance on statistical significance to judge study findings, it also inflates effect size estimates if there are more small studies (as is usually the case in education). Smaller studies need larger effect sizes to reach statistical significance. The omission of negative or null findings and a preponderance of smaller studies together with reliance on statistical significance all produce a positive bias.
Random-effects model:	Statistical model used to combine or pool effect sizes from the included studies. In this model (compared with the *fixed-effect model*) the variability in the effects of an intervention or approach in the studies included in the meta-analysis is assumed to depend on two things, the within-study variance (like the fixed-effect model) *and* the between-studies variance. This assumes that each study has been sampled from a distribution

of effects. Each study is assigned a weight based on these two factors. The model assumes a different underlying effect for each study and takes this into consideration in combining the effect sizes. This produces a more conservative estimate statistically but may inflate the overall estimate if there is publication bias resulting from inclusion of a number of small studies with larger positive effects.

Randomised
controlled
trial (RCT):
This is where two or more groups have been formed through random allocation (or a similar method). This is the only method which ensures that selection bias is eliminated at baseline. Both known *and* unknown *co-variates* which might influence outcomes are likely to equally distributed among the groups.

Sample size
calculations:
Trials in educational research commonly exhibit a Type II error. This is where the sample size is insufficient to show, as statistically significant, a difference that is educationally important. Reviews of educational interventions have shown that most interventions will, at best, only lead to an improvement in the region of half a standard deviation and quite often somewhat less. Statistical theory shows that to reliably detect (with 80% power) half a standard deviation difference as statistically significant (p = 0.05) for a normally distributed variable requires a minimum sample size of 126 participants between the two arms. Studies that are smaller than this risk erroneously concluding that there was not a significant difference when actually there was. Therefore, a good quality study ought to describe the reasoning behind the choice of sample size.

Search strategy:
Strategy used to identify primary studies to be included in a meta-analysis.

Standard deviation:
A measure of spread or dispersion of continuous data. A high standard deviation implies that the values are widely scattered relative to the mean value, whilst a small value implies the converse.

Statistical significance:	In traditional hypothesis testing, a result is said to have statistical significance when it is considered unlikely to have occurred, given the null hypothesis. In education the level of likelihood is usually set at the 95% level (or one in twenty). The *p*-value of a result is the probability of obtaining a result at least as extreme, assuming that the null hypothesis were true and that the sample is randomly selected from a given population. The term significance does not imply importance in this context. The approach was developed by Ronald Fisher in the last century, but is problematic in the way that it is now applied and used (see the discussion in Chapters 1 and Chapter 4).
Subgroup analysis:	Statistical analysis used to test the effect of a categorical moderator.
Systematic review:	A review where explicit definitions, criteria and the search strategy are specified and use to identify studies for inclusion in a review. It aims to address selection bias reviewing.
Vote counting:	This approach draws conclusions about a specific area of study by comparing the number of studies in which there are statistically significant results with the number of studies reporting non-significant results or negative results. It has two major problems: First, it is highly susceptible to publication bias; and second, it is more likely to reach a false negative conclusion, particularly if effect sizes in a field are small.
Weight:	In a meta-analysis, when study effect sizes are combined, a weight is assigned to each study. The weight assigned is based on the inverse of the variance and its precise value depends on the model chosen for combining the effect sizes (see *fixed-effect* and *random-effects* models).

References

References marked with * are included in Appendix A, providing more details about each specific meta-analysis and the moderator variables exploring variation in effects.

Abrami*, P. C., Bernard, R. M., Borokhovski, E., Wade, A., Surkes, M. A., Tamim, R., & Zhang, D. (2008). Instructional interventions affecting critical thinking skills and dispositions: A stage 1 meta-analysis. *Review of Educational Research* 78 (4), 1102–1134. http://dx.doi.org/10.3102 /0034654308326084

Aguinis, H., Pierce, C. A., Bosco, F. A., Dalton, D. R., & Dalton, C. M. (2011). Debunking myths and urban legends about meta-analysis. *Organizational Research Methods*, 14 (2), 306–331. https://doi.org/10.1177/1094428110375720

Ahn, S., Ames, A. J., & Myers, N. D. (2012). A review of meta-analyses in education: Methodological strengths and weaknesses. *Review of Educational Research*, 82 (4), 436–476. http://dx.doi.org/10.3102/0034654312458162

Anwar, E., Goldberg, E., Fraser, A., Acosta, C. J., Paul, M., & Leibovici, L. (2014). Vaccines for preventing typhoid fever. *The Cochrane Library*. http://dx .doi.org/10.1002/14651858.CD001261.pub3

Arnold, R. D. (1968). Four methods of teaching word recognition to disabled readers. *The Elementary School Journal*, 68 (5), 269–274. http://dx.doi.org/10 .1086/460445

Athappilly, K., Smidchens, U., & Kofel, J. W. (1983). A computer-based meta-analysis of the effects of modern mathematics in comparison with traditional mathematics. *Educational Evaluation and Policy Analysis*, 5 (4), 485–493. http://www.jstor.org/stable/1164053

Baigent, C., Blackwell, L., Emberson, J., Holland, L. E., Reith, C., Bhala, N., & Collins, R. (2010). Efficacy and safety of more intensive lowering of LDL cholesterol: A meta-analysis of data from 170,000 participants in 26 randomised trials. *The Lancet*, 376 (9753), 1670–1681.

Bangert-Drowns*, R. L., Hurley, M. M., & Wilkinson, B. (2004). The effects of school-based writing-to-learn interventions on academic achievement: A meta-analysis. *Review of Educational Research*, 74 (1), 29–58. https://doi.org /10.3102/00346543074001029

Bangert-Drowns*, R. L., Kulik, C. L. C., Kulik, J. A., & Morgan, M. (1991). The instructional effect of feedback in test-like events. *Review of Educational Research*, 61 (2), 213–238. http://dx.doi.org/10.3102/00346543061002213

Bernstein, B. (1971) *Class, codes and control: Theoretical studies towards a sociology of language*. London: Routledge & Kegan Paul.

Bayraktar*, S. (2000). A meta-analysis of the effectiveness of computer assisted instruction in science education. *Journal of Research on Technology in Education*, 42 (2), 173–188. http://www.dx.doi.org/10.1080/15391523.2001.10782344

Bennett, R. E. (2011). Formative assessment: A critical review. *Assessment in Education: Principles, Policy & Practice*, 18 (1), 5–25. http://dx.doi.org/10.1080/0969594X.2010.513678

Berkeley, S., Scruggs, T. E., & Mastropieri, M. A. (2010). Reading comprehension instruction for students with learning disabilities, 1995–2006: A meta-analysis. *Remedial and Special Education*, 31 (6), 423–436. http://dx.doi.org/10.1177/0741932509355988

Berninger, V. W., Vaughan, K., Abbott, R. D., Begay, K., Coleman, K. B., Curtin, G., & Graham, S. (2002). Teaching spelling and composition alone and together: Implications for the simple view of writing. *Journal of Educational Psychology*, 94 (2), 291–304. http://dx.doi.org/10.1037/0022–0663.94.2.291

Biesta, G. (2007). Why 'what works' won't work: Evidence-based practice and the democratic deficit in educational research. *Educational Theory*, 57 (1), 1–22. http://dx.doi.org/10.1111/j.1741–5446.2006.00241.x

Black P., & Wiliam, D. (1998). Assessment and classroom learning, *Assessment in Education*, 5, 7–73. http://dx.doi.org/10.1080/0969595980050102

Black, P., & Wiliam, D. (2005). Lessons from around the world: How policies, politics and cultures constrain and afford assessment practices. *Curriculum Journal*, 16, 249–261. http://dx.doi.org/10.1080/09585170500136218

Black, P., & Wiliam, D. (2009). Developing the theory of formative assessment. *Educational Assessment, Evaluation and Accountability*, 21 (1), 5–31. http://dx.doi.org/10.1007/s11092-008-9068-5

Blok*, H., Oostdam, R., Otter, M. E., & Overmaat, M. (2002). Computer-assisted instruction in support of beginning reading instruction: A review. *Review of Educational Research*, 72 (1), 101–130. http://dx.doi.org/10.3102/00346543072001101

Bloom, B. S. (1984). The 2 sigma problem: The search for methods of group instruction as effective as one-to-one tutoring. *Educational Researcher*, 13 (6), 4–16. http://dx.doi.org/10.3102/0013189X013006004

Bloom, B. S., Hastings, J. T., & Madaus, G. F. (Eds.) (1971). *Handbook on the formative and summative evaluation of student learning*. New York: McGraw-Hill.

Bloom, B. S. (1964). *Stability and change in human characteristics*. New York: Wiley.

Borenstein, M., Hedges, L. V., Higgins, J., & Rothstein, H. R. (2009). *Introduction to meta-analysis*. London: John Wiley & Sons, Ltd.

Box, G. E., Hunter, W. G., & Hunter, J. S. (1978). *Statistics for experimenters: An introduction to design, data analysis, and model building*. New York: Wiley.

Bracht, G. H., & Glass, G. V. (1968). The external validity of experiments. *American Educational Research Journal*, 5 (4), 437–474. https://doi.org/10.3102/00028312005004437

Briggs, D. C. (2016). Can Campbell's law be mitigated? In H. Braun (Ed.), *Meeting the challenges to measurement in an era of accountability* (pp. 168–179). New York: Routledge.

Bus*, A. G., Van Ijzendoorn, M. H., & Pellegrini, A. D. (1995). Joint book reading makes for success in learning to read: A meta-analysis on intergenerational transmission of literacy. *Review of Educational Research*, 65 (1), 1–21. http://dx .doi.org/10.3102/00346543065001001

Camilli, G., Vargas, S., & Yurecko, M. (2003). Teaching children to read: The fragile link between science and federal education policy. *Education Policy Analysis Archives*, 11 (15), n15. http://dx.doi.org/10.14507/epaa .v11n15.2003

Camilli*, G., Vargas, S., Ryan, S., & Barnett, W. S. (2008). Meta-Analysis of the effects of early education interventions on cognitive and social development. *Teachers College Record*, 112 (3), 579–620. http://dx.doi.org/10.14507/epaa .v11n15.2003

Campbell, D. T. (1976). Assessing the Impact of Planned Social Change. Occasional Paper Series, # 8. Kalamazoo, MI: Western Michigan University Evaluation Centre.

Campbell, M. K., Piaggio, G., Elbourne, D. R., & Altman, D. G. (2012). Consort 2010 statement: Extension to cluster randomised trials. *British Medical Journal*, 345, e5661. http://dx.doi.org/10.1136/bmj.e5661

Chall, J. S. (1983). *Learning to read: The great debate*. New York: McGraw-Hill.

Chalmers, I., & Altman, D. G. (1995) *Systematic Reviews*. London: BMJ Publications.

Chalmers, I., Hedges L. V., & Cooper, H. (2002). A brief history of research synthesis. *Evaluation and the Health Professions* 25, 12–37. http://dx.doi.org/10 .1037/0033–2909.87.3.442

Chan, M. E., & Arvey, R. D. (2012). Meta-analysis and the development of knowledge. *Perspectives on Psychological Science*, 7 (1), 79–92. http://dx.doi.org/10 .1177/1745691611429355

Chauhan*, S. (2017). A meta-analysis of the impact of technology on learning effectiveness of elementary students. *Computers & Education*, 105, 14–30. http:// dx.doi.org/10.1016/j.compedu.2016.11.005

Cheung*, A. C., & Slavin, R. E. (2012). How features of educational technology applications affect student reading outcomes: A meta-analysis. *Educational Research Review*, 7 (3), 198–215. http://doi.org/10.1016/j.edurev.2012.05.002

Cheung*, A. C., & Slavin, R. E. (2013). The effectiveness of educational technology applications for enhancing mathematics achievement in K-12 class-rooms: A meta-analysis. *Educational Research Review*, 9, 88–113. http://doi.org/10 .1016/j.edurev.2013.01.001

Cheung, A. C., & Slavin, R. E. (2015). How methodological features affect effect sizes in education. In *Best Evidence Encyclopedia*. Baltimore: Johns Hopkins University. http://www.bestevidence.org/word/methodologi cal_Sept_21_2015.pdf

Childs, A., & Menter I. (Eds.) *Mobilising teacher researchers: Challenging educational inequality*. London: Routledge.

Chiu*, C. W. T. (1998). Synthesizing metacognitive interventions: What training characteristics can improve reading performance? Paper presented at the Annual Meeting of the American Educational Research Association, San Diego, CA, April 13–17, 1998. http://files.eric.ed.gov/fulltext/ED420844.pdf

Churches, R. (2016). *Closing the gap: Test and learn.* Nottingham: National College for Teaching and Leadership.

Cipriani, A., Higgins, J. P., Geddes, J. R., & Salanti, G. (2013). Conceptual and technical challenges in network meta-analysis. *Annals of Internal Medicine,* 159 (2), 130–137. http://dx.doi.org/10.7326/0003-4819-159-2-201307160 -00008

Clark*, D. B., Tanner-Smith, E. E., & Killingsworth, S. S. (2016). Digital games, design, and learning: A systematic review and meta-analysis. *Review of Educational Research,* 86 (1), 79–122. http://dx.doi.org/10.3102/0034654315582065

Coe, R. (2002). It's the effect size stupid; what effect size is and why is it important. Paper presented at the Annual Conference of the British Educational Research Association, University of Exeter, England, September 12–14, 2002.

Coffield, F., Moseley, D., Hall, E., & Ecclestone, K. (2004). *Learning styles and pedagogy in post-16 learning. A systematic and critical review.* London: Learning and Skills Research Centre.

Cohen, J. (1988). *Statistical power analysis for the behavioral sciences.* (2nd ed.) Hilsdale, NJ: Lawrence Erlbaum Associates.

Coldwell, M., Greany, T., Higgins, S., Brown, C., Maxwell, B., Stiell, B., Stoll, L., Willis, B., & Helen Burns, H. (2017) *Evidence-informed teaching: An evaluation of progress in England* (Research Report July 2017 (DFE-RR-696)). London: Department for Education. Retrieved from: https://www.gov .uk/government/uploads/system/uploads/attachment_data/file/625007/Eviden ce-informed_teaching_-_an_evaluation_of_progress_in_England.pdf

Comfort*, C. B. (2003). *Evaluating the effectiveness of parent training to improve outcomes for young children: A meta-analytic review of the published research.* (Ph.D. in Applied Psychology). University of Calgary. http://dspace.ucalgary .ca/handle/1880/42284

Cooper, H. M., & Rosenthal, R. (1980). Statistical versus traditional procedures for summarizing research findings. *Psychological Bulletin,* 87 (3), 442. http://dx .doi.org/10.1037/0033-2909.87.3.442

Corak, M. (2013). Income inequality, equality of opportunity, and intergenera-tional mobility. *The Journal of Economic Perspectives,* 27 (3), 79–102. http://hdl .handle.net/10419/80702

Cordingley, P. (2008). Research and evidence-informed practice: Focusing on practice and practitioners. *Cambridge Journal of Education,* 38 (1), 37–52. http:// dx.doi.org/10.1080/03057640801889964

Cronbach, L. J., Ambron, S. R., Dornbusch, S. M., Hess, R. O., Hornik, R. C., Phillips, D. C., Walker, D. F., & Weiner, S. S. (1980). *Toward reform of program evaluation: Aims, methods, and institutional arrangements.* San Francisco, CA: Jossey-Bass.

Cuevas, J. (2015). Is learning styles-based instruction effective? A comprehensive analysis of recent research on learning styles. *Theory and Research in Education,* 13 (3), 308–333. http://dx.doi.org/10.1177/1477878515606621

Cummings, C., Laing, K., Law, J., McLaughlin, J., Papps, I., Todd, L., & Woolner, P. (2012). *Can changing aspirations and attitudes impact on educational attainment? A review of interventions.* York: Joseph Rowntree Foundation.

Dagenais, C., Lysenko, L., Abrami, P. C., Bernard, R. M., Ramde, J., & Janosz, M. (2012). Use of research-based information by school practitioners and determinants of use: A review of empirical research. *Evidence & Policy: A Journal of Research, Debate and Practice*, 8 (3), 285–309. http://dx.doi.org/10 .1332/174426412X654031

D'Angelo*, C., Rutstein, D., Harris, C., Bernard, R., Borokhovski, E., & Haertel, G. (2014). *Simulations for STEM learning: Systematic review and meta-analysis*. Menlo Park, CA: SRI International. www.sri.com/education

Davies, F. W. J. (1973). *Teaching reading in early England*. London: Pitman and Sons, Ltd.

Davis*, D. S. (2010). *A meta-analysis of comprehension strategy instruction for upper elementary and middle school students* (Doctoral dissertation). Vanderbilt University, USA). http://etd.library.vanderbilt.edu/available/etd-06162010 –100830/unrestricted/Davis_dissertation.pdf

Deaton, A., & Cartwright, N. (2016). *Understanding and misunderstanding randomized controlled trials* (Working Paper No. 22595). Cambridge MA: National Bureau of Economic Research.

De Boer, H., Donker, A. S., & van der Werf, M. P. (2014). Effects of the attributes of educational interventions on students' academic performance: A meta-analysis. *Review of Educational Research*, 84 (4), 509–545. http://dx .doi.org/10.3102/0034654314540006

Department for Education (2012). *What is the research evidence on writing?* (Research Report DFE-RR238). London: Department for Education. https:// www.gov.uk/government/uploads/system/ . . . /DFE-RR238.pdf

DerSimonian, R., & Laird, N. (1986). Meta-analysis in clinical trials. *Controlled Clinical Trials*, 7 (3), 177–188. https://doi.org/10.1016/0197-2456 (86) 90046-2

Dignath*, C., Buettner, G., & Langfeldt, H. (2008). How can primary school students learn self-regulated learning strategies most effectively? A meta-analysis on self-regulation training programmes. *Educational Research Review* 3 (2), 101–129. http://www.dx.doi.org/10.1016/j.edurev.2008.02.003

Dillon, J.T., (1982). Superanalysis. *American Journal of Evaluation* 3 (3), 35–43.

Donker*, A. S., De Boer, H., Kostons, D., Dignath van Ewijk, C. C., & Van der Werf, M. P. C. (2014). Effectiveness of learning strategy instruction on academic performance: A meta-analysis. *Educational Research Review*, 11, 1–26. http://www.dx.doi.org/10.1016/j.edurev.2013.11.002

Duffin, J., & Simpson, A. (2000). Understanding their thinking: The tension between the cognitive and the affective. *In Perspectives on adults learning mathematics* (pp. 83–99). Springer Netherlands.

Durlak, J. A., & DuPre, E. P. (2008). Implementation matters: A review of research on the influence of implementation on program outcomes and the factors affecting implementation. *American Journal of Community Psychology*, 41 (3–4), 327–350. http://dx.doi.org/10.1007/s10464-008-9165-0

Edmonds*, M. S., Vaughn, S., Wexler, J., Reutebuch, C., Cable, A., Tackett, K. K., & Schnakenberg, J. W. (2009). A synthesis of reading interventions and effects on reading comprehension outcomes for older struggling readers. *Review of Educational Research*, 79(1), 262–300. http://dx.doi.org/10 .3102/0034654308325998

EEF (2017). *Improving literacy in key stage two: Guidance report.* London: Education Endowment Foundation.

Ehri*, C. L., Nunes, S.R., Stahl, S. A., & Willows, D. M. (2001). Systematic phonics instruction helps students learn to read: Evidence from the National Reading Panel's meta-analysis. *Review of Educational Research,* 71 (3), 393–447. http://dx.doi.org/10.3102/00346543071003393

Elleman*, A. M., Lindo, E. J., Morphy, P., & Compton, D. L. (2009). The impact of vocabulary instruction on passage-level comprehension of school-age children: A meta-analysis. *Journal of Research on Educational Effectiveness,* 2 (1), 1–44. https://doi.org/10.1080/19345740802539200

Elliott, J. H., Turner, T., Clavisi, O., Thomas, J., Higgins, J. P., Mavergames, C., & Gruen, R. L. (2014). Living systematic reviews: An emerging opportunity to narrow the evidence-practice gap. *PLoS Medicine,* 11 (2), e1001603. http://dx.doi.org/10.1371/journal.pmed.1001603

Elwood, P. C., Cochrane, A. L., Burr, M. L., Sweetnam, P. M., Williams, G., Welsby, E., Hughes, S .J., & Renton, R. (1974). A randomized controlled trial of acetyl salicylic acid in the secondary prevention of mortality from myocardial infarction. *British Medical Journal,* 1(5905), 436.

Epstein, J. L. (2009). *School, family, and community partnerships: Your handbook for action* (3rd ed.).Thousand Oaks, CA: Corwin Press.

Erlenmeyer-Kimling, L., & Jarvik, L. F. (1963). Genetics and intelligence: A review. *Science,* 142(3598), 1477–1479. http://dx.doi.org/10.1126/science .142.3598.1477

Eysenck, H. J. (1952). The effects of psychotherapy: An evaluation. *Journal of Consulting Psychology,* 1 (5), 319.

Eysenck, H. J. (1978). An exercise in mega-silliness. *American Psychologist,* 33 (5), 517. http://dx.doi.org/10.1037/0003-066X.33.5.517.a

Fan, X., & Chen, M. (2001). Parental involvement and students' academic achievement: A meta-analysis. *Educational Psychology Review,* 13 (1), 1–22. http://dx.doi.org/10.1023/A:1009048817385

Fauzan*, N. (2003). *The effects of metacognitive strategies on reading comprehension: A quantitative synthesis and the empirical investigation* (Doctoral dissertation). University of Durham). http://etheses.dur.ac.uk/1086/

Fisher, R. A. (1935) *The design of experiments.* Edinburgh: Oliver and Boyd.

Fisher, R. A. (1956) *Statistical methods and scientific inference.* Edinburgh: Oliver and Boyd.

Fitzgerald, J., & Shanahan, T. (2000) Reading and writing relations and their development, *Educational Psychologist,* 35 (1), 39–50. http://dx.doi.org/10 .1207/S15326985EP3501_5

Fitz-Gibbon, C. T. (1984). Meta-analysis: An explication. *British Educational Research Journal,* 10 (2), 135–144. http://dx.doi.org/10.1080/0141192840100202

Fitz-Gibbon, C. T. (1985). The implications of meta-analysis for educational research. *British Educational Research Journal,* 11 (1), 45–49. http://dx.doi.org /10.1080/0141192850110105

Flesch, R. (1955). *Why Johnnie can't read: And what you can do about it.* New York: Harper & Brothers.

Francis, G. (2012). Too good to be true: Publication bias in two prominent studies from experimental psychology. *Psychonomic Bulletin & Review*, 19 (2), 151–156. http://dx.doi.org/10.3758/s13423-012-0227-9

Fraser, A., Paul, M., Goldberg, E., Acosta, C. J., & Leibovici, L. (2007). Typhoid fever vaccines: Systematic review and meta-analysis of randomised controlled trials. *Vaccine*, 25 (45), 7848–7857. http://dx.doi.org/10.1016/j.vaccine.2007 .08.027

Fraser, B. J., Walberg, H. J., Welch, W. W., & Hattie, J. A. (1987). Syntheses of educational productivity research. *International Journal of Educational Research*, 11 (2), 147–252. http://dx.doi.org/10.1016/0883–0355(87)90035–8

Fuchs*, L. S., & Fuchs, D. (1986). Effects of systematic formative evaluation: A meta-analysis. *Exceptional Children*, 53 (3), 199–208. http://dx.doi.org/10 .1177/001440298605300301

Fukkink*, R. G., & De Glopper, K. (1998). Effects of instruction in deriving word meaning from context: A meta-analysis. *Review of Educational Research*, 68 (4), 450–469. http://www.dx.doi.org/10.3102/00346543068004450

Furukawa, T. A., & Leucht, S. (2011) How to obtain NNT from Cohen's d: Comparison of two methods. *PloS One*, 6 (4), e19070. http://dx.doi.org/10 .1371/journal.pone.0019070

Galloway, A. M. (2003). *Improving reading comprehension through metacognitive strategy instruction: Evaluating the evidence for the effectiveness of the reciprocal teaching procedure* (Doctoral dissertation ETD collection). Lincoln, NE: University of Nebraska–Lincoln. AAI3092542. http://digitalcommons.unl .edu/dissertations/AAI3092542

Galuschka*, K., Ise, E., Krick, K., & Schulte-Körne, G. (2014) Effectiveness of treatment approaches for children and adolescents with reading disabilities: A meta-analysis of randomized controlled trials. *PLoS One*, 9 (2), e89900. http:// dx.doi.org/10.1371/journal.pone.0089900

Garlinger*, D. K., & Frank, B. M. (1986). Teacher-student cognitive style and academic achievement: A review and a mini-meta analysis. *Journal of Classroom Interaction*, 21 (2), 2–8. http://www.jstor.org/stable/23869505

Glass, G. V. (1976). Primary, secondary, and meta-analysis of research. *Educational Researcher*, 5 (10), 3–8. http://dx.doi.org/10.3102/0013189X005010003

Glass, G. V. (1977). Integrating findings: The meta-analysis of research. *Review of Research in Education*, 5 (1), 351–379. http://dx.doi.org/10.3102 /0091732X005001351

Glass, G. V. (2000). Meta-analysis at 25. Retrieved from: http://www.gvglass.info /papers/meta25.html

Glass, G. V, McGaw, B., & Smith, M. L. (1981). *Meta-analysis in social research.* Beverly Hills, CA: Sage Publications.

Goldacre, B. (2010). *Bad science: Quacks, hacks, and big pharma flacks.* Toronto: McClelland & Stewart.

Goldacre, B. (2014). *Bad pharma: How drug companies mislead doctors and harm patients.* London: Macmillan.

Goldberg, A., Russell, M., & Cook, A. (2003). The effect of computers on student writing: A meta-analysis of studies from 1992 to 2002. *The Journal of Technology, Learning and Assessment*, 2 (1). https://ejournals.bc.edu/ojs/index .php/jtla/article/view/1661

Goodwin, A. P., & Ahn, S. (2010). A meta-analysis of morphological interventions: Effects on literacy achievement of children with literacy difficulties. *Annals of Dyslexia*, 60 (2), 183–208. http://dx.doi.org/10.1080/10888438.2012.689791

Gorard, S. (2014). The widespread abuse of statistics by researchers: What is the problem and what is the ethical way forward? *Psychology of Education Review* 38 (1), 3–10.

Gorard, S., & See, B. H. (2013). *Do parental involvement interventions increase attainment? A review of the evidence.* London: Nuffield Foundation. http://www.nuffieldfoundation.org/sites/default/files/files/Do_parental_involvement_interventions_increase_attainment1.pdf

Gorard, S., See, B. H., & Davies, P. (2012). *The impact of attitudes and aspirations on educational attainment and participation.* York: Joseph Rowntree Foundation.

Gorard, S., See, B. H., & Siddiqi, N. (2015) Philosophy for children (P4C) evaluation. (Report). London: EEF

Gough, P. B., & Tunmer, W. E. (1986). Decoding, reading, and reading disability. *Remedial and Special Education*, 7, 6–10. http://dx.doi.org/10.1177/074193258600700104

Graham, S., & Hebert, M. A. (2010). *Writing to read: Evidence for how writing can improve reading* (A Carnegie Corporation Time to Act Report). Washington, DC: Alliance for Excellent Education.

Graham, S., Hebert, M., & Harris, K. R. (2015). Formative assessment and writing. *The Elementary School Journal*, 115 (4), 523–547. http://dx.doi.org/10.1086/681947

Graham*, S., McKeown, D., Kiuhara, S., & Harris, K. R. (2012) A meta-analysis of writing instruction for students in the elementary grades. *Journal of Educational Psychology.* 104 (4), 879–896. http://dx.doi.org/10.1037/a0029185

Graham, S., & Perrin, D. (2007). A meta-analysis of writing instruction for adolescent students. *Journal of Educational Psychology*, 99 (3), 445. http://dx.doi.org/10.1037/0022–0663.99.3.445

Graham, S., Liu, X., Aitken, A., Ng, C., Bartlett, B., Harris, K. R., & Holzapfel, J. (2017). Effectiveness of literacy programs balancing reading and writing instruction: A meta-analysis. reading research quarterly (online early). Retrieved from: http://www.dx.doi.org/10.1002/rrq.194

Grant, M. J., & Booth, A. (2009). A typology of reviews: An analysis of 14 review types and associated methodologies. *Health Information & Libraries Journal*, 26 (2), 91–108. http://dx.doi.org/ 10.1111/j.1471–1842.2009.00848.x

Griffiths, Y. M., & Snowling, M. J. (2002). Predictors of exception word and nonword reading in dyslexic children: The severity hypothesis. *Journal of Educational Psychology*, 94 (1), 34–43. http://dx.doi.org/10.1037/0022–0663.94.1.34

Guthrie, J. T., McRae, A., & Klauda, S. L. (2007). Contributions of concept-oriented reading instruction to knowledge about interventions for motivations in reading. *Educational Psychologist*, 42 (4), 237–250. http://www.dx.doi.org/10.1080/00461520701621087

Guzetti*, B. J., Snyder, T. E., Glass, G. V., & Gamas, W. S. (1993). Meta-analysis of instructional interventions from reading education and science education to promote conceptual change in science. *Reading Research Quarterly,* 28 (2), 116–161.

Haller*, E. P., Child, D. A., & Walberg, H. J. (1988). Can comprehension be taught? A quantitative synthesis of 'metacognitive studies'. *Educational Researcher,* 17 (9), 5–8. http://www.dx.doi.org/10.3102/0013189X017009005

Harris, K. R., Graham, S., & Mason, L. H. (2006) Improving the writing, knowledge, and motivation of struggling young writers: Effects of self-regulated strategy development with and without peer support. *American Educational Research Journal,* 43 (2), 295–340. http://dx.doi.org/10.3102/00028312043002295

Hattie, J. (1992). Measuring the effects of schooling. *Australian Journal of Education,* 36 (1), 5–13. http://dx.doi.org/10.1177/000494419203600102

Hattie, J. A. (2008). *Visible Learning.* London: Routledge.

Hattie, J.A. (2011). *Visible Learning for teachers.* London: Routledge.

Hattie, J. (2015). The applicability of Visible Learning to higher education. *Scholarship of Teaching and Learning in Psychology,* 1 (1), 79. http://dx.doi.org/10.1037/stl0000021

Hattie, J., & Timperley, H. (2007). The power of feedback. *Review of Educational Research* 77 (1), 81–112. http://dx.doi.org/10.3102/003465430298487

Hedges, L. V. (1983). A random effects model for effect sizes. *Psychological Bulletin,* 93(2), 388. http://dx.doi.org/10.1037/0033–2909.93.2.388

Hedges, L. V., & Olkin, I. (1980). Vote-counting methods in research synthesis. *Psychological Bulletin,* 88 (2), 359–369. http://dx.doi.org/10.1037/0033–2909.88.2.359

Hemsley-Brown, J. V., & Sharp, C. (2004). The use of research to improve professional practice: A systematic review of the literature. *Oxford Review of Education,* 29 (4), 449–470. http://dx.doi.org/10.1080/0305498032000153025

Higgins, S. (2003). Parlez-vous mathematics? In I. Thompson (Ed.), *Enhancing primary mathematics teaching and learning.* Buckingham: Open University Press.

Higgins, S. E. (2013). Matching style of learning. In J. Hattie & E. M. Anderman (Eds.) *International guide to student achievement (educational psychology handbook)* (pp. 337–438). London: Routledge.

Higgins, S. (2016). Meta-synthesis and comparative meta-analysis of education research findings: Some risks and benefits. *Review of Education,* 4 (1), 31–53. http://dx.doi.org/10.1002/rev3.3067

Higgins, S. (2017). Room in the Toolbox? The place of randomised controlled trials in educational research. In A. Childs & I. Menter (Eds.) *Mobilising teacher researchers: Challenging educational inequality.* London: Routledge.

Higgins, S., & Hall, E. (2004). Picking the strawberries out of the jam: Thinking critically about narrative reviews, systematic reviews and meta-analysis. Presented at the British Education Research Association conference, at Manchester Metropolitan University, September 2004. Retrieved from: http://www.leeds.ac.uk/educol/documents/00003835.htm

Higgins, S., & Katsipataki, M. (2015). Evidence from meta-analysis about parental involvement in education which supports their children's learning. *Journal of Children's Services,* 10 (3), 1–11. http://dx.doi.org/10.1108/JCS-02-2015-0009

Higgins, S., & Katsipataki, M. (2016). Communicating comparative findings from meta-analysis in educational research: Some examples and suggestions. *International Journal of Research & Method in Education*, 39 (3), 237–254 http://dx.doi.org/10.1080/1743727X.2016.1166486

Higgins, S., Katsipataki, M., Coleman, R., Henderson, P., Major, L. E., Coe, R., & Mason, D. (2016). *The Sutton Trust-Education Endowment Foundation Teaching and Learning Toolkit: Oral language interventions*. London: Education Endowment Foundation. https://educationendowmentfoundation.org.uk/tool kit/toolkit-a-z/oral-language-interventions/

Higgins, S., Katsipataki, M., Kokotsaki, D., Coleman, R., Major, L. E., & Coe, R. (2014). *The Sutton Trust-Education Endowment Foundation Teaching and Learning Toolkit*. London: Education Endowment Foundation. http://edu cationendowmentfoundation.org.uk/toolkit/

Higgins, S., Katsipataki, M., Kokotsaki, D., Coleman, R., Major, L. E., & Coe, R. (2013). The Sutton Trust-Education Endowment Foundation Teaching and Learning Toolkit: Technical Appendices London: Education Endowment Foundation. Retrieved from: http://educationendowmentfounda tion.org.uk/uploads/pdf/Technical Appendices (June_2013).pdf

Higgins, S., Kokotsaki, D., & Coe, R. (2011) *Toolkit of strategies to improve learning: Summary for schools spending the Pupil Premium*. London: Sutton Trust.

Higgins, S., Wall, K., Baumfield, V., Hall, E., Leat, D., Moseley, D., & Woolner, P. (2007). *Learning to learn in schools phase 3 evaluation*. Newcastle: Newcastle University.

Higgins*, S., Hall, E., Baumfield, V., & Moseley, D. (2005). A meta-analysis of the impact of the implementation of thinking skills approaches on pupils. In *Research Evidence in Education Library*. London: EPPI-Centre, Social Science Research Unit, Institute of Education, University of London. http://e ppi.ioe.ac.uk/cms/Default.aspx?tabid=339

Higgins, S., & Simpson, A. (2011). Visible Learning: A synthesis of over 800 meta-analyses relating to achievement. By John AC Hattie: Book review. *British Journal of Educational Studies*, 59 (2), 197–201. http://dx.doi.org/10.1080/00071005.2011.584660

Hill, C. J., Bloom, H. S., Black, A. R., & Lipsey, M. W. (2008). Empirical benchmarks for interpreting effect sizes in research. *Child Development Perspectives*, 2(3), 172–177. http://dx.doi.org/10.1111/j.1750–8606.2008.00061.x

Hill*, N. E., & Tyson, D. F. (2009). Parental involvement in middle school: A meta-analytic assessment of the strategies that promote achievement. *Developmental Psychology*, 45 (3), 740. http://dx.doi.org/10.1037/a0015362

Hunt, M. (1997). *How science takes stock: The story of meta-analysis*. New York: Russell Sage Foundation.

Ioannidis, J. P. (2009). Integration of evidence from multiple meta-analyses: A primer on umbrella reviews, treatment networks and multiple treatments meta-analyses. *Canadian Medical Association Journal*, 181 (8), 488–493. http://dx.doi.org/10.1503/cmaj.081086

Ioannidis, J. P. A., & Trikalinos, T. A. (2007). The appropriateness of asymmetry tests for publication bias in meta-analyses: A large survey. *Canadian Medical Association Journal*, 176, 8. http://dx.doi.org/10.1503/cmaj.060410

Jacob, R., & Parkinson, J. (2015). The potential for school-based interventions that target executive function to improve academic achievement A review. *Review of Educational Research*. Retrieved from: http://dx.doi.org/10.3102 /0034654314561338

Jeynes*, W. (2012). A meta-analysis of the efficacy of different types of parental involvement programs for urban students. *Urban Education*, 47 (4), 706–742. http://dx.doi.org/10.1177/0042085912445643

Jeynes*, W. H. (2005). A meta-analysis of the relation of parental involvement to urban elementary school student academic achievement. *Urban Education*, 40 (3), 237–269. http://dx.doi.org/10.1177/0042085905274540

Jeynes*, W. H. (2007). The relationship between parental involvement and urban secondary school student academic achievement a meta-analysis. *Urban Education*, 42 (1), 82–110. http://dx.doi.org/10.1177/00420859062 93818

Jeynes*, W. H. (2008). A meta-analysis of the relationship between phonics instruction and minority elementary school student academic achievement. *Education and Urban Society*, 40 (2), 151–166. http://dx.doi.org/10.1177 /0013124507304128

Kanadli*, S. (2016). A meta-analysis on the effect of instructional designs based on the learning styles models on academic achievement, attitude and retention. *Educational Sciences: Theory and Practice*, 16 (6), 2057–2086. http://dx.doi.org /10.12738/estp.2016.6.0084

Kao, G., & Thompson J. S. (2003). Racial and ethnic stratification in educational achievement and attainment. *Annual Review of Sociology*, 29, 417–442. http:// dx.doi.org/10.1146/annurev.soc.29.010202.100019

Katsipataki, M., & Higgins, S. (2016). What works or what's worked? Evidence from education in the United Kingdom. *Procedia-Social and Behavioral Sciences*, 217, 903–909. http://dx.doi.org/10.1016/j.sbspro.2016.02.030

Kavale, K. A., & LeFever, G. B. (2007). Dunn and Dunn model of learning-style preferences: Critique of Lovelace meta-analysis. *The Journal of Educational Research*, 101 (2), 94–97. http://dx.doi.org/10.3200/JOER.101.2.94–98

Kavale, K., Hirshoren, A., & Forness, S. (1998). Meta-analytic validation of the Dunn-and-Dunn model of learning-style preferences: A critique of what was Dunn. *Learning Disabilities Research and Practice*, 13, 75–80.

Kavale*, K. A., & Forness, S. R. (1987). Substance over style: Assessing the efficacy of modality testing and teaching. *Exceptional Children*, 54 (3), 228–39. http://journals.sagepub.com/doi/pdf/10.1177/001440298705400305

Kazrin, A., Durac, J., & Agteros, T. (1979). Meta-meta analysis: A new method for evaluating therapy outcome. *Behaviour Research and Therapy*, 17 (4), 397–399. http://dx.doi.org/10.1016/0005-7967(79)90011-1

Kim*, J. S., & Quinn, D. M. (2013). The effects of summer reading on low-income children's literacy achievement from kindergarten to grade 8 a meta-analysis of classroom and home interventions. *Review of Educational Research*, 83 (3), 386–431. http://dx.doi.org/10.3102/0034654313483906

Kingston*, N., & Nash, B. (2011). Formative assessment: A meta-analysis and call for research. *Educational Measurement: Issues and Practice*, 30 (4), 28–37. http://dx.doi.org/10.1111/j.1745-3992.2011.00220.x

Klauer*, K. J., & Phye, G. D. (2008). Inductive reasoning: A training approach. *Review of Educational Research*, 78 (1), 85–123. http://www.dx.doi.org/10.3102/0034654307313402

Klein, P. D. (2003) Rethinking the multiplicity of cognitive resources and curricular representations: Alternatives to 'learning styles' and 'multiple intelligences. *Journal of Curriculum Studies*, 35 (1), 45–81. http://dx.doi.org/10.1080/00220270210141891

Kluger*, A. N., & DeNisi, A. (1996). The effects of feedback interventions on performance: A historical review, a meta-analysis, and a preliminary feedback intervention theory. *Psychological Bulletin*, 119 (2), 254. http://dx.doi.org/10.1037/0033-2909.119.2.254

Komarraju, M., Karau, S. J., Schmeck, R. R., & Avdic, A. (2011). The Big Five personality traits, learning styles, and academic achievement. *Personality and Individual Differences*, 51(4), 472–477. https://doi.org/10.1016/j.paid.2011.04.019

Kulik*, J. A., & Fletcher, J. D. (2016). Effectiveness of intelligent tutoring systems: A meta-analytic review. *Review of Educational Research*, 86 (1), 42–78. https://doi.org/10.3102/0034654315581420

Kulik, C., Kulik, J., & Bangert-Drowns, R. (1990). Effectiveness of mastery learning programs: A meta-analysis. *Review of Educational Research*, 60 (2), 265–306. http://dx.doi.org/10.3102/00346543060002265

Kulik, J. A., & Kulik, C. L. C. (1989). The concept of meta-analysis. *International Journal of Educational Research*, 13 (3), 227–340. http://dx.doi.org/10.1016/0883-0355(89)90052-9

Kulik*, J. A., & Fletcher, J. D. (2016). Effectiveness of intelligent tutoring systems: A meta-analytic review. *Review of Educational Research*, 86 (1), 42–78. http://dx.doi.org/10.3102/0034654315581420

Kunkel*, A. K. (2015). *The effects of computer-assisted instruction in reading: A meta-analysis* (Doctoral dissertation). University of Minnesota. http://conservancy.umn.edu/handle/11299/175221

Langer, L., Tripney, J., & Gough, D. (2016). *The science of using science: Researching the use of research evidence in decision-making*. London: EPPI-Centre, Social Science Research Unit, UCL Institute of Education, University College London.

Lather, P. (1986). Research as praxis. *Harvard Educational Review*, 56 (3), 257–278.

Layzer*, J. I., Goodson, B. D., Bernstein, L., & Price, C. (2001). *National evaluation of family support programs. Final report volume A: The meta-analysis* (ED462186). Washington, DC: Administration for Children, Youth, and Families (DHHS). http://files.eric.ed.gov/fulltext/ED462186.pdf

Leak, J., Duncan, G. J., Li, W., Magnuson, K., Schindler, H., & Yoshikawa, H. (2013). *Is timing everything? How early childhood education program impacts vary by starting age, program duration and time since the end of the program* (Working Paper). National Forum on Early Childhood Policy and Programs, Meta-analytic Database Project. Center on the Developing Child, Harvard University.

Lemons, C. J., Fuchs, D., Gilbert, J. K., & Fuchs, L. S. (2014). Evidence-based practices in a changing world: Reconsidering the counterfactual in education research. *Educational Researcher,* 43 (5), 242–252. http://dx.doi.org/10.3102/0013189X14539189

Levin, B. (2011).Mobilising research knowledge in education. *London Review of Education,* 9(1), 15–26. http://dx.doi.org/10.1080/14748460.2011.550431

Lewis*, R. J., & Vosburgh, W. T. (1988). Effectiveness of kindergarten intervention programs: A meta-analysis. *School Psychology International,* 9 (4), 265–275. http://dx.doi.org/10.1177/0143034388094004

Li*, Q., & Ma, X. (2010). A meta-analysis of the effects of computer technology on school students' mathematics learning. *Educational Psychology Review,* 22 (3), 215–243. http://dx.doi.org/10.1007/s10648-010-9125-8

Lipsey, M. W., & Wilson, D. (2000). *Practical meta-analysis (applied social research methods).* London: Sage Publications.

Losinski*, M., Cuenca-Carlino, Y., Zablocki, M., & Teagarden, J. (2014). Examining the efficacy of self-regulated strategy development for students with emotional or behavioral disorders: A meta-analysis. *Behavioral Disorders,* 40 (1), 52–67. http://dx.doi.org/10.17988/0198-7429-40.1.52

Lou*, Y., Abrami, P. C., & d'Apollonia, S. (2001). Small group and individual learning with technology: A meta-analysis. *Review of Educational Research,* 71 (3), 449–521. http://dx.doi.org/10.3102/00346543071003449

Lovelace, M. K. (2005). Meta-analysis of experimental research based on the Dunn and Dunn model. *The Journal of Educational Research,* 98 (3), 176–183. http://dx.doi.org/10.3200/JOER.98.3.176–183

Lovelace*, M. K. (2002). *A meta-analysis of experimental research studies based on the Dunn and Dunn learning-style model* (ProQuest Dissertations and Theses, p. 177). New York: School of Education and Human Services, St. John's University. Retrieved from: http://search.proquest.com/docview/275698679

Luyten, H. (2006). An empirical assessment of the absolute effect of schooling: Regression/discontinuity applied to TIMSS-95. *Oxford Review of Education,* 32 (3), 397–429. http://dx.doi.org/10.1080/03054980600776589

Lysakowski*, R. S., & Walberg, H. J. (1982). Instructional effects of cues, participation, and corrective feedback: A quantitative synthesis. *American Educational Research Journal,* 19 (4), 559–578. http://dx.doi.org/10.3102/00028312019004559

Makel, M. C., & Plucker, J. A. (2014). Facts are more important than novelty: Replication in the education sciences. *Educational Researcher,* 43 (6), 304–316. http://dx.doi.org/10.3102/0013189X14545513

Manning*, M., Homel, R., & Smith, C. (2010). A meta-analysis of the effects of early developmental prevention programs in at-risk populations on non-health outcomes in adolescence. *Children and Youth Services Review,* 32 (4), 506–519. http://dx.doi.org/10.1016/j.childyouth.2009.11.003

Manz*, P. H., Hughes, C., Barnabas, E., Bracaliello, C., & Ginsburg-Block, M. (2010). A descriptive review and meta-analysis of family-based emergent literacy interventions: To what extent is the research applicable to low-income, ethnic-minority or linguistically-diverse young children? *Early Childhood Research Quarterly,* 25 (4), 409–431. http://dx.doi.org/10.1016/j.ecresq.2010.03.002

Marulis*, L. M., & Neuman, S. B. (2010). The effects of vocabulary intervention on young children's word learning: A meta-analysis. *Review of*

Educational Research, 80 (3), 300–335. http://www.dx.doi.org/10.3102/0034654310377087

Marzano, R. J. (1998). *A theory-based meta-analysis of research on instruction.* Aurora, CO: Mid-Continent Regional Educational Lab (McREL).

Mayer, R. E. (2011). Does styles research have useful implications for educational practice? *Learning and Individual Differences*, 21, 319–320. http://dx.doi.org/10.1016/j.lindif.2010.11.016

McArthur*, G., Eve, P. M., Jones, K., Banales, E., Kohnen, S., Anandakumar, T., Larsen L., Marinus, E., Wang, H. C., & Castles, A. (2012). Phonics training for English-speaking poor readers. *Cochrane Database of Systematic Reviews*, 12, CD009115. http://dx.doi.org/10.1002/14651858.CD009115.pub2

McKenna, M. C., & Stahl, K. A. D. (2015). *Assessment for reading instruction* (3rd ed.). New York: Guilford Publications.

Means*, B., Toyama, Y., Murphy, R., Bakia, M., & Jones, K. (2009). *Evaluation of evidence-based practices in online learning: A meta-analysis and review of online learning studies.* US Department of Education. http://files.eric.ed.gov/fulltext/ED505824.pdf

Melby-Lervåg, M., Lyster, S. A. H., & Hulme, C. (2012). Phonological skills and their role in learning to read: A meta-analytic review. *Psychological Bulletin*, 138 (2), 322. http://dx.doi.org/10.1037/a0026744

Mills, E. J., Thorlund, K., & Ioannidis, J. P. (2013). Demystifying trial networks and network meta-analysis. *British Medical Journal*, 346, f2914. http://dx.doi.org/10.1136/bmj.f2914

Moher, D., Liberati, A., Tetzlaff, J., & Altman, D. G. (2009). Preferred reporting items for systematic reviews and meta-analyses: The PRISMA statement. *Annals of Internal Medicine*, 151 (4), 264–269. http://dx.doi.org/10.7326/0003-4819-151-4-200908180-00135

Mol, S. E., Bus, A. G., & de Jong, M. T. (2009). Interactive book reading in early education: A tool to stimulate print knowledge as well as oral language. *Review of Educational Research*, 79 (2), 979–1007. https://doi.org/10.3102/0034654309332561

Moran*, J., Ferdig, R. E., Pearson, P. D., Wardrop, J., & Blomeyer, R. L. (2008). Technology and reading performance in the middle-school grades: A meta-analysis with recommendations for policy and practice. *Journal of Literacy Research*, 40 (1), 6–58. http://www.dx.doi.org/10.1080/10862960802070483

Morphy*, P., & Graham, S. (2012). Word processing programs and weaker writers/readers: A meta-analysis of research findings, *Reading and Writing*, 25, 641–678. http://dx.doi.org/10.1007/s11145-010-9292-5

Moseley, D., Baumfield, V., Elliott, J., Higgins, S., Miller, J., & Newton, D. P. (2005). *Frameworks for thinking: A handbook for teaching and learning.* Cambridge: Cambridge University Press.

Nickerson, R. S. (2000). Null hypothesis significance testing: A review of an old and continuing controversy. *Psychological Methods*, 5 (2), 241–301. http://dx.doi.org/10.1037/1082-989X.5.2.241

Nye*, C., Schwartz, J., & Turner, H. (2006). Approaches to parent involvement for improving the academic performance of elementary school age children:

A systematic review. *Campbell Systematic Reviews*, 2 (4). http://campbellcolla boration.org/lib/download/63/

O'Rourke, K. (2007). An historical perspective on meta-analysis: Dealing quantitatively with varying study results. *Journal of the Royal Society of Medicine*, 100 (12), 579–582. http://dx.doi.org/10.1177/0141076807100012020

Olejnik, S., & Algina, J. (2000). Measures of effect size for comparative studies: Applications, interpretations, and limitations. *Contemporary Educational Psychology*, 25 (3), 241–286. http://dx.doi.org/10.1006/ceps.2000.1040

Onuoha*, C. O. (2007). *Meta-analysis of the effectiveness of computer-based laboratory versus traditional hands-on laboratory in college and pre-college science instructions* (Order No. 3251334). ProQuest Dissertations & Theses Global. (304699656). Retrieved from: http://search.proquest.com/docview/304699656

Open Science Collaboration. (2015). Estimating the reproducibility of psychological science. *Science*, 349 (6251), aac4716. http://dx.doi.org/10.1126/science .aac4716

Pashler, H., McDaniel, M., Rohrer, D., & Bjork, R. (2008). Learning styles: Concepts and evidence. *Psychological Science in the Public Interest*, 9 (3), 106–119. http://dx.doi.org/10.1111/j.1539–6053.2009.01038.x

Pearson, K. (1904) Report on certain enteric fever inoculation statistics. *The British Medical Journal*, 2 (2288), 1243–1246. http://www.jstor.org/stable/ 20282622

Pearson, P. D., & Dole, J. A. (1987). Explicit comprehension instruction: A review of research and a new conceptualization of instruction. *The Elementary School Journal*, 88 (2), 151–165. http://dx.doi.org/10.1086 /461530

Pearson*, P. D., Ferdig, R. E., Blomeyer Jr., R. L., & Moran, J. (2005). *The effects of technology on reading performance in the middle-school grades: A meta-analysis with recommendations for policy.* Oak Brook, IL: Learning Point Associates/North Central Regional Educational Laboratory (NCREL).

Perry, V., Albeg, L., & Tung, C. (2012). Meta-analysis of single-case design research on self-regulatory interventions for academic performance. *Journal of Behavioral Education*, 21 (3), 217–229. http://www.dx.doi.org/10.1007/s1086 4-012–9156-y

Peirce, C. S., & Jastrow, J. (1885) On small differences in sensation. *Memoirs of the National Academy of Sciences*, 3, 73–83. https://philpapers.org/archive/PEIOSD .pdf

Pratt, J. G., Smith, B. M., Rhine, J. B., Stuart, C. E., & Greenwood, J. A. (1940). *Extra-sensory perception after sixty years: A critical appraisal of the research in extra-sensory perception.* New York: Henry Holt and Company. http://dx.doi.org/10 .1037/13598–000

Protzko, J. (2015). The environment in raising early intelligence: A meta-analysis of the fadeout effect. *Intelligence*, 53, 202–210. http://dx.doi.org/10.1016/j.intell .2015.10.006

Raudenbush, S. W. (1997). Statistical analysis and optimal design for cluster randomized trials. *Psychological Methods*, 2 (2), 173. http://dx.doi.org/10.1037 /1082-989X.2.2.173

Roberts, R. W., & Coleman, J. C. (1958). An investigation of the role of visual and kinesthetic factors in reading failure, *The Journal of Educational Research*, 51 (6), 445–451. http://dx.doi.org/10.1080/00220671.1958.10882487

Rosen*, Y., & Salomon, G. (2007). The differential learning achievements of constructivist technology-intensive learning environments as compared with traditional ones: A meta-analysis. *Journal of Educational Computing Research*, 36 (1), 1–14. https://doi.org/10.2190/R8M4-7762-282U-554J

Rosenthal, R. (1966) *Experimenter effects in behavioral research*. New York: Appleton-Century-Crofts.

Rosenzweig, C. (2001). *A meta-analysis of parenting and school success: The role of parents in promoting students' academic performance*. Paper presented at the Annual Meeting of the American Educational Research Association (Seattle, WA, April 10–14, 2001). ED452232 http://files.eric.ed.gov/full text/ED452232.pdf

Savage, R., Burgos, G., Wood, E., & Piquette, N. (2015). The simple view of reading as a framework for national literacy initiatives: A hierarchical model of pupil-level and classroom-level factors. *British Educational Research Journal*, 41 (5), 820–844. http://dx.doi.org/10.1002/berj.3177

Schunk, D. H. (2008). Metacognition, self-regulation, and self-regulated learning: Research recommendations. *Educational Psychology Review*, 20 (4), 463–467. http://www.dx.doi.org/10.1007/s10648-008-9086-3

Seidel, T., & Shavelson, R. J. (2007). Teaching effectiveness research in the past decade: The role of theory and research design in disentangling meta-analysis results. *Review of Educational Research*, 77 (4), 454–499. http://dx.doi.org/10 .3102/0034654307310317

Sellke, T., Bayarri, M. J., & Berger, J. O. (2001). Calibration of p values for testing precise null hypotheses. *The American Statistician*, 55 (1), 62–71. http://dx.doi .org/10.1198/000313001300339950

Sénéchal*, M., & Young, L. (2008). The effect of family literacy interventions on children's acquisition of reading from kindergarten to grade 3: A meta-analytic review. *Review of Educational Research*, 78 (4), 880–907. http://dx.doi.org/10 .3102/0034654308320319

Seo*, Y. J., & Bryant, D. P. (2009). Analysis of studies of the effects of computer-assisted instruction on the mathematics performance of students with learning disabilities. *Computers & Education*, 53 (3), 913–928. http://dx .doi.org/10.1016/j.compedu.2009.05.002

Scammacca*, N. K., Roberts, G., Vaughn, S., & Stuebing, K. K. (2015). A meta-analysis of interventions for struggling readers in Grades 4–12: 1980–2011. *Journal of Learning Disabilities*, 48 (4), 369–390. http://www.dx.doi.org/10 .1177/0022219413504995

Schagen, I., & Hodgen, E. (2009). *How much difference does it make? Notes on understanding, using, and calculating effect sizes for schools*. Wellington: NCZER www.educationcounts.govt.nz/publications/schooling/36097/36098

Sheldon, S. B. (2009), Improving student outcomes with school, family, and community partnerships: A research review. In J. L. Epstein et al. (Eds.) *School, family, and community partnerships: Your handbook for action*, 3rd ed. (pp. 40–56).Thousand Oaks, CA: Corwin Press.

Sherman*, K. H. (2007). *A meta-analysis of interventions for phonemic awareness and phonics instruction for delayed older readers* (Doctoral thesis UMI No. 3285626). University of Oregon. ProQuest Dissertations and Theses. Retrieved from: http://search.proquest.com/docview/304825094

Simpson, A. (2017). The misdirection of public policy: Comparing and combining standardized effect sizes. *Journal of Education Policy*, 32 (4), 450–466. https://doi .org/10.1080/02680939.2017.1280183

Sipe, T. A., & Curlette, W. L. (1996) .A meta-synthesis of factors related to educational achievement: A methodological approach to summarizing and synthesizing meta-analyses. *International Journal of Educational Research*, 25 (7), 583–698. https://doi.org/10.1016/S0883–0355(96)80001–2

Slavin, R. E. (1986). Best-evidence synthesis: An alternative to meta-analytic and traditional reviews. *Educational Researcher*, 15 (9), 5–11. http://dx.doi.org/10 .3102/0013189X015009005

Slavin, R. E. (2002). Evidence-based education policies: Transforming educational practice and research. *Educational Researcher*, 31 (7), 15–21. http://dx.doi .org/10.3102/0013189X031007015

Slavin, R. E., Lake, C., Davis, S., & Madden, N. A. (2011). Effective programs for struggling readers: A best-evidence synthesis. *Educational Research Review*, 6 (1), 1–26. https://doi.org/10.1016/j.edurev.2010.07.002

Slavin, R., & Madden, N. A. (2011). Measures inherent to treatments in program effectiveness reviews. *Journal of Research on Educational Effectiveness*, 4 (4), 370–380. http://dx.doi.org/10.1080/19345747.2011.558986

Slavin*, R. E., Lake, C., Davis, S., & Madden, N. A. (2011). Effective programs for struggling readers: A best-evidence synthesis. *Educational Research Review*, 6 (1), 1–26. http://dx.doi.org/10.1016/j.edurev.2010.07.002

Slavin, R., & Smith, D. (2009). The relationship between sample sizes and effect sizes in systematic reviews in education. *Educational Evaluation and Policy Analysis*, 31 (4), 500–506. http://dx.doi.org/10.3102/0162373709352369

Slemmer*, D. L. (2002). *The effect of learning styles on student achievement in various hypertext, hypermedia and technology enhanced learning environments: A meta-analysis* (Ph.D. dissertation, unpublished). Boise, ID: Boise State University (ProQuest Dissertations and Theses). https://www.editlib.org/p/122865/

Smith, E., & Gorard, S. (2005). They don't give us our marks: The role of formative feedback in student progress. *Assessment in Education* 12 (1), 21–38. http://dx.doi.org/10.1080/0969594042000333896

Smith, N. L. (1982). Evaluative applications of meta-and mega-analysis. *American Journal of Evaluation*, 34, 43–47. http://dx.doi.org/10.1177/1098214 08200300412

Smith, M. L., & Glass, G. V. (1977). Meta-analysis of psychotherapy outcome studies. *American Psychologist*, 32 (9), 752.

Snook, I., O'Neill, J., Clark, J., O'Neill, A. M., & Openshaw, R. (2009). Invisible learnings? A commentary on John Hattie's book: Visible Learning: A synthesis of over 800 meta-analyses relating to achievement. *New Zealand Journal of Educational Studies*, 44 (1), 93.

Speight, S., Callanan, M., Griggs, J., & Farias, J. (2016). *Rochdale research into practice: Evaluation report and executive summary*. London: EEF.

Stahl, S. A., & Miller, P. D. (1989). Whole language and language experience approaches for beginning reading: A quantitative research synthesis. *Review of Educational Research*, 59 (1), 87–116. http://dx.doi.org/10.3102/0034654305900 1087

Steenbergen-Hu*, S., & Cooper, H. (2013). A meta-analysis of the effectiveness of intelligent tutoring systems on K–12 students' mathematical learning. *Journal of Educational Psychology*, 105 (4), 970–987. http://dx.doi.org/10.1037 /a0032447

Stone, C. A. (1998). The metaphor of scaffolding: Its utility for the field of learning disabilities. *Journal of Learning Disabilities*, 31 (4), 344–364. http://dx .doi.org/10.1177/002221949803100404

Strong*, G. K., Torgerson, C. J., Torgerson, D., & Hulme, C. (2011). A systematic meta- analytic review of evidence for the effectiveness of the 'Fast ForWord' language intervention program. *Journal of Child Psychology and Psychiatry*, 52 (3), 224–235. http://www.dx.doi.org/10.1111/j.1469–7610 .2010.02329.x

Susser, M. (1977). Judgment and causal inference: Criteria in epidemiologic studies. *American Journal of Epidemiology*, 105 (1), 1–15. http://dx.doi.org/10 .1093/oxfordjournals.aje.a112349

Swanson, E., Vaughn, S., Wanzek, J., Petscher, Y., Heckert, J., Cavanaugh, C., & Tackett, K. (2011). A synthesis of read-aloud interventions on early reading outcomes among preschool through third graders at risk for reading difficulties. *Journal of Learning Disabilities*, 44 (3), 258–275. http://www.dx.doi.org/10 .1177/0022219410378444

Sweet, M. A., & Appelbaum, M. I. (2004). Is home visiting an effective strategy? A meta-analytic review of home visiting programs for families with young children. *Child development*, 75 (5), 1435–1456. http://dx.doi.org/10.1111/j .1467–8624.2004.00750.x

Tamir*, P. (1985). Meta-analysis of cognitive preferences and learning. *Journal of Research in Science Teaching*, 22 (1), 1–17. http://dx.doi.org/10.1002/tea .3660220101

Tenenbaum*, G., & Goldring, E. (1989). A meta-analysis of the effect of enhanced instruction: Cues, participation, reinforcement and feedback, and correctives on motor skill learning. *Journal of Research and Development in Education*, 22 (3), 53–64.

Terhart, E. (2011). Has John Hattie really found the Holy Grail of research on teaching? An extended review of Visible Learning. *Journal of Curriculum Studies*, 43 (3), 425–438.

Tingir*, S., Cavlazoglu, B., Caliskan, O., Koklu, O., & Intepe-Tingir, S. (2017). Effects of mobile devices on K–12 students' achievement: A meta-analysis. *Journal of Computer Assisted Learning (early view)*. Retrieved from: http://dx.doi .org/10.1111/jcal.12184

Todd, E. S., & Higgins, S. (1998). Powerlessness in professional and parent partnerships. *British Journal of Sociology of Education*, 19 (2), 227–236. http://dx .doi.org/10.1080/0142569980190205

Tokpah*, C. L. (2008). *The effects of computer algebra systems on students' achievement in mathematics* (Order No. 3321336). ProQuest Dissertations & Theses

Global. (304549974). Retrieved from: http://search.proquest.com/docview/304549974

Torgerson, D., Torgerson, C., Ainsworth, H., Buckley, H. M., Heaps, C. K., Hewitt, C., & Mitchell, N. (2014). *Improving writing quality: Evaluation report and executive summary May 2014*. London: EEF. http://educationendowment foundation.org.uk/uploads/pdf/EEF_Evaluation_Report_-_Improving_Writing_ Quality_-_May_2014_v2.pdf

Torgerson*, C., & Zhu, D. (2003). A systematic review and meta-analysis of the effectiveness of ICT on literacy learning in English, 5–16. In *Research Evidence in Education Library*. London: EPPI-Centre, Social Science Research Unit, Institute of Education.

Torgerson*, C., Brooks, G., & Hall, J. (2006). *A systematic review of the research literature on the use of phonics in the teaching of reading and spelling* (DfES Research Report RR711). London: DfES Publications.

Torgerson*, C. J., & Elbourne, D. (2002). A systematic review and meta-analysis of the effectiveness of information and communication technology (ICT) on the teaching of spelling. *Journal of Research in Reading*, 25, 129–143. http://www .dx.doi.org/10.1111/1467-9817.00164

Tracy, B., Reid, R., & Graham, S. (2009). Teaching young students strategies for planning and drafting stories: The impact of self-regulated strategy development. *The Journal of Educational Research*, 102 (5), 323–332. http://dx .doi.org/10.3200/JOER.102.5.323-332

Tunnell, M. O., & Jacobs, J. S. (1989). Using 'real' books: Research findings on literature based reading instruction. *The Reading Teacher*, 42 (7), 470–477. http://www.jstor.org/stable/20200193

Tymms, P. (2004). Are standards rising in English primary schools? *British Educational Research Journal*, 30 (4), 477–494. http://dx.doi.org/10.1080 /0141192042000237194

Umbach, B., Darch, C., & Halpin, G. (1989). Teaching reading to low performing first graders in rural schools: A comparison of two instructional approaches. *Journal of Instructional Psychology*, 16 (3), 112.

Underwood, B. J. (1957). Interference and forgetting. *Psychological Review*, 64 (1), 49. http://dx.doi.org/10.1037/h0044616

Valentine, J. C., Pigott, T. D., & Rothstein, H. R. (2010). How many studies do you need? A primer on statistical power for meta-analysis. *Journal of Educational and Behavioral Statistics*, 35 (2), 215–247. http://dx.doi.org/10.3102 /1076998609346961

van de Ven, M., Voeten, M., Steenbeek-Planting, E. G., & Verhoeven, L. (2017). Post-primary reading fluency development: A latent change approach. *Learning and Individual Differences*, 55, 1–12. http://dx.doi.org/10.1016/j.lindif.2017.02.001

Van Steensel*, R., McElvany, N., Kurvers, J., & Herppich, S. (2011). How effective are family literacy programs? Results of a meta-analysis. *Review of Educational Research*, 81 (1), 69–96. http://dx.doi.org/10.3102/0034654310388819

Van Voorhis, F. L., Maier, M. F., Epstein, J. L., Lloyd, C. M., & Leuong, T. (2013). *The impact of family involvement on the education of children ages 3 to 8: A focus on literacy and math achievement outcomes and social-emotional skills*. New York: Center on School, Family and Community Partnerships, MDRC.

Walberg, H. J. (1982). Educational productivity: Theory, evidence, and prospects. *Australian Journal of Education*, 26 (2), 115–122. http://dx.doi.org/10.1177/000494418202600202

Wang, M. C., Haertel, G. D., & Walberg, H. J. (1993). Toward a knowledge base for school learning. *Review of Educational Research*, 63 (3), 249–294. https://doi.org/10.3102/00346543063003249

Waxman, H. C., Lin, M-F., &Michko, G. M. (2002). A meta-analysis of the effectiveness of teaching and learning with technology on student outcomes. Napier, IL: Learning Point Associates.

Wigelsworth, M., Lendrum, A., Oldfield, J., Scott, A., ten Bokkel, I., Tate, K., & Emery, C. (2016). The impact of trial stage, developer involvement and international transferability on universal social and emotional learning programme outcomes: A meta-analysis. *Cambridge Journal of Education*, 46 (3), 347–376. http://dx.doi.org/10.1080/0305764X.2016.1195791

Wiliam, D. (2010). Standardized testing and school accountability. *Educational Psychologist*, 45 (2), 107–122. http://dx.doi.org/10.1080/00461521003703060

Wiliam, D., Lee, C., Harrison, C., & Black, P. (2004). Teachers developing assessment for learning: Impact on student achievement. *Assessment in Education: Principles, Policy & Practice*, 11 (1), 49–65. http://dx.doi.org/10.1080/0969594042000208994

Wiseman, S. (Ed.). (1961). *Examinations and English education*. Manchester, UK: Manchester University Press.

Wouters*, P., Van Nimwegen, C., Van Oostendorp, H., & Van Der Spek, E. D. (2013). A meta-analysis of the cognitive and motivational effects of serious games. *Journal of Educational Psychology*, 105 (2), 249–265. http://dx.doi.org/10.1037/a0031311

Wyse, D., & Styles, M. (2007). Synthetic phonics and the teaching of reading: The debate surrounding England's 'Rose Report'. *Literacy*, 41 (1), 35–42. http://dx.doi.org/10.1111/j.1467-9345.2007.00455.x

Xiao, Z., Kasim, A., & Higgins, S. E. (2016) Same difference? Understanding variation in the estimation of effect sizes from educational trials. *International Journal of Educational Research* 77, 1–14 http://dx.doi.org/10.1016/j.ijer.2016.02.001

Yusuf, S., Peto, R., Lewis, J., Collins, R., & Sleight, P. (1985). Beta blockade during and after myocardial infarction: An overview of the randomized trials. *Progress in Cardiovascular Diseases*, 27 (5), 335–371. http://dx.doi.org/10.1016/S0033-6620(85)80003-7

Zheng*, L. (2016). The effectiveness of self-regulated learning scaffolds on academic performance in computer-based learning environments: A meta-analysis. *Asia Pacific Education Review*, 17 (2), 187–202. http://dx.doi.org/10.1007/s12564-016-9426-9

Index